# Tracer Diffusion Data
# for Metals, Alloys, and Simple Oxides

# Tracer Diffusion Data for Metals, Alloys, and Simple Oxides

## John Askill

*Physics Department*
*Millikin University, Decatur, Illinois*

IFI/Plenum • New York—Washington—London • 1970

*Library of Congress Catalog Card Number 73-95202*
SBN 306-65147-5


© 1970 IFI/Plenum Data Corporation
A Subsidiary of Plenum Publishing Corporation
227 West 17th Street, New York, N.Y. 10011

United Kingdom edition published by Plenum Press, London
A Division of Plenum Publishing Company, Ltd.
Donington House, 30 Norfolk Street, London W.C.2, England

Printed in the United States of America

# Preface

Atomic diffusion in metals was first discovered some sixty-five years ago, and since then a considerable wealth of data has accumulated on diffusion in various systems. However, work prior to about the year 1940 is now mainly of historical interest, since experiments were often carried out under experimental conditions and with methods of analysis leading to uncertainties in interpreting the measured diffusion coefficients. Data on diffusion rates are of importance in processes which are controlled by rates of atomic migration such as growth of phases and homogenization of alloys. In addition diffusion plays an important part in theories of such phenomena as oxidation, plastic deformation, sintering, and creep.

A tremendous advance in diffusion studies was made possible by the availability of radioactive isotopes of sufficiently high specific activity after the second world war. Measurements of self-diffusion rates then became possible using radioactive isotopes having the same chemical properties as the solvent material, and it also became possible to study tracer impurity diffusion when the concentration of the impurity is so small as not to alter the chemical homogeneity of the system. In the last ten to fifteen years the purity of materials used in diffusion studies has increased considerably and the methods of analysis have become more standardized.

The main purpose of this book is to bring together in one compilation all the radioactive tracer diffusion data of metals in pure metals, alloys, and simple oxides that have been reported in the literature between 1938 and December 1968.

The  data  have  been  divided  into  four  parts:

I.    Self-diffusion in pure  metals.
II.   Impurity diffusion in pure metals.
III.  Self- and impurity diffusion in alloys.
IV.   Self- and impurity diffusion in simple metal oxides.

Over 1200 diffusion entries are included with elements and solutes
arranged alphabetically.  A complete list of references and an in-
dex of authors are also given.

The various aspects of the diffusion data listed in the tables,
such as temperature range, method of analysis of the diffusion
samples, and the diffusion parameters $D_0$ and $Q$ in the Arrhenius
equation $D = D_0 \exp(-Q/RT)$ are discussed in an introductory chapter. The various mechanisms which have been proposed for atomic diffusion in solids are also described..

Often one needs values of the diffusion coefficient for a particular system which has not yet been studied experimentally.  In this
case one or more of the many empirical relations between diffusion and other physical properties of solids may be useful.  For
this reason, the various empirical relations for diffusion, particularly self-diffusion, are discussed in a second introductory
chapter. In this way the applied scientist can make a reasonable
estimate of the diffusion coefficient at some temperature in most
systems for which direct experimental data are not available.

Most of the work involved in compiling these tables was carried out while the author was at the University of Reading, Reading, England, and at the Oak Ridge National Laboratory, Oak Ridge,
Tennessee.

# Contents

*Chapter I*

# Diffusion, the Diffusion Coefficient, and Mechanisms of Diffusion

## 1. DIFFUSION

Diffusion is the way in which matter is transported through matter. It occurs by approximately random motions of the atoms in a crystal lattice. The net result of many such random movements of a large number of atoms is actual displacement of matter, the movement being activated by the thermal energy of the crystal. In a pure material any particular atom is continually moving from one position to another in the material. This is called s e l f - d i f f u s i o n and can be studied experimentally by the use of radioactive tracers.

## 2. THE DIFFUSION COEFFICIENT

The transfer of heat also occurs by random motions of atoms or molecules. This led Fick[1] to produce what is now generally known as Fick's first law of diffusion,

$$J = -D\frac{dc}{dx} \tag{1.1}$$

where $J$ is the diffusion flux or the rate of transfer of atoms of a particular constituent of a system through unit area in unit time, $dc/dx$ is the concentration gradient in the direction of diffusion, and $D$ is the diffusion coefficient. The units of $J$ are atoms/cm$^2$ · sec, and those of c atoms/cm$^3$, so that $D$ is measured in cm$^2$/sec.

---

[1]A. Fick, Ann. Phys. Leipzig, 94:59 (1855).

Equation (1.1) is similar to that for heat flow where heat flux is directly proportional to the temperature gradient, and also to the electrical conductivity analogue. If the fluxes of atoms through two planes a distance dx apart are J and $J + dJ/dc$, then the difference in flux is

$$\frac{dJ}{dc} = \left(-D\frac{dc}{dx}\right) - \left(-D\frac{dc}{dx} + \frac{d}{dx}D\frac{dc}{dx}\right) = -\frac{d}{dx}\left(D\frac{dc}{dx}\right)$$

Since no atoms are produced or lost (a continuity condition), this difference in flux must be equal to $-dc/dt$. This gives us Fick's second law of diffusion:

$$\frac{dc}{dt} = \frac{d}{dx}\left(D\frac{dc}{dx}\right) \tag{1.2}$$

For self-diffusion and impurity diffusion at tracer concentrations, the diffusion coefficient D is independent of concentration c so that equation (1.2) becomes

$$\frac{dc}{dt} = D\frac{d^2c}{dx^2} \tag{1.3}$$

Diffusion experiments are analyzed in terms of a suitable solution of equation (1.2) or (1.3) for the particular geometrical conditions used.

## 3. SOLUTIONS TO THE DIFFUSION EQUATION

Both the initial conditions and the boundary conditions of the system have to be determined. The initial conditions fix the state before diffusion takes place (t = 0). The boundary conditions contain geometric factors defining the experimental conditions such as specimen dimensions and continuity conditions.

The majority of diffusion experiments are carried out with a semi-infinite specimen with an infinitesimally thin layer of radioactive isotope deposited on one face (Fig. 1A). In this case the solution of the general diffusion equation (1.3) is

$$c(x, t) = \frac{S}{\sqrt{\pi D_T t}}\ \exp\left(-\frac{x^2}{4D_T t}\right) \tag{1.4}$$

where $c(x, t)$ is the concentration of radioactive atoms at a distance x from the initial face, S is the initial total amount of activity, t is the time of the diffusion anneal, and $D_T$ is the diffusion coefficient

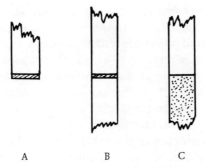

                    A              B            C

Fig. 1. Experimental arrangement of samples. (A)
Semi-infinite specimen — infinitesimally thin layer
of isotope; (B) double semi-infinite specimen — infinitesimally thin layer of isotope; (C) semi-infinite specimen — semi-infinite uniformly active
specimen.

at a temperature of $T°K$. Equation (1.4) can be simplified to

$$c(x) = c(0) \exp\left(-\frac{x^2}{4D_T t}\right) \tag{1.5}$$

since S and t are constants for a particular diffusion sample obtained from a particular diffusion anneal.

Sometimes two semi-infinite specimens with an infinitesimally
thin layer of radioactive material between them are used (Fig. 1B).
In this case the solution of the diffusion equation is

$$c(x, t) = \frac{S}{2\sqrt{\pi D_T t}} \exp\left(-\frac{x^2}{4D_T t}\right) \tag{1.6}$$

This is of the same form as for a single sample, i.e.,

$$c(x) = c(0) \exp\left(-\frac{x^2}{4Dt}\right)$$

One other arrangement that is sometimes used is a semi-
infinite specimen with a semi-infinite uniformly active second
specimen (Fig. 1C). In this case the concentration distribution
equation can be shown to be

$$c(x) = \frac{c(0)}{2}\left[1 - \text{erf}\left(\frac{x}{2\sqrt{Dt}}\right)\right] \tag{1.7}$$

The aim of any method of analysis is to measure the concentration $c(x)$ as a function of distance x through the diffused specimen. If the concentration distribution obeys equation (1.5), for example, then a plot of $\log_e c(x)$ versus $x^2$ will be linear with slope $-1/4D_T t$, giving a value of the diffusion coefficient $D_T$ at the particular temperature T°K of the anneal. The four most widely used methods of analysis are the serial sectioning method, the residual activity method, the surface decrease method, and autoradiography. Nearly all of the entries in the bibliography of diffusion data given in Chapter 3 used one or more of these four methods of analysis. Each will be discussed here in some detail.

## 4. METHODS OF ANALYSIS OF THE DIFFUSION SAMPLES

### 4.1. Serial Sectioning

The serial sectioning method is the method of analysis most frequently used in the study of diffusion using radioactive tracers. The diffusion specimen is subjected to a series of sectioning operations, the specimen being weighed before and after each cut in order to determine the thickness of material removed. The cuttings are collected and their activity measured with a Geiger tube or a scintillation counter. From the weights the coordinates of the midpoints of each slice (x) are calculated, and a plot of $\log_e$ (count rate) versus $x^2$ then has a slope of $-1/4D_T t$ for volume diffusion. Hence, if the time of the anneal t is known, the diffusion coefficient $D_T$ may be calculated. If the log (count rate) versus $x^2$ relationship is not linear over the full length of the penetration plot, this means that bulk diffusion is not the only mechanism present. Grain boundary diffusion is usually observed at low temperatures in polycrystalline specimens where a linear log (count rate) versus x plot is expected. Upward curvature at the high $x^2$ end of the log (count rate) versus $x^2$ plot is then normally associated with grain boundary or dislocation diffusion.

The main advantage of the serial sectioning method is that it is simple, direct, and does not depend on the properties of the radiations from the radioactive material used. In addition, it does not presuppose the nature of the diffusion phenomenon. If the values of $\sqrt{2Dt}$ are between $10^{-5}$ and $10^{-2}$ cm, the sectioning is conveniently performed on a precision lathe. The smallest value of the diffusion coefficient that can conveniently be measured by lathe sectioning is about $10^{-11}$ cm$^2$/sec. To make it possible to measure smaller diffusion coefficients, the method can be modified by removing the slices by grinding. If a microbalance is available to measure the material removed to about 1 $\mu$g, diffusion coefficients down to $10^{-13}$ to $10^{-14}$ cm$^2$/sec can be measured in this way. Recently a process of anodizing and stripping has been developed[2] by which diffusion coefficients as low as $10^{-6}$ to $10^{-17}$ cm$^2$/sec can be measured. With this process, which at present is only applicable to niobium and tungsten, very small layers of the order of 100 Å can be removed. Serial sectioning is regarded as the most reliable and most accurate method of diffusion analysis.

## 4.2. Residual Activity Method

The residual activity (or Gruzin) method is similar to the serial sectioning analysis in that layers of thickness x are removed by lathe sectioning, grinding, etching, or anodizing, but the total remaining activity of the sample I is measured in this case. If one considers an elementary thickness dx between x and x + dx, we have

$$dI = k \frac{S}{\sqrt{\pi D_T t}} \exp\left(-\frac{x^2}{4D_T t}\right) \exp\left[-\mu(x - x_n)\right]$$

where $\mu$ is the absorption coefficient for the radiation by the solvent material, $x_n$ is the thickness of the n-th layer, $I_n$ is the activity of the sample after n layers are removed, and k is a constant depending on the geometric counting conditions. Integrating this equation gives

$$I_n = k \frac{S}{\sqrt{\pi D_T t}} \int_{x_n}^{\infty} \exp\left(-\frac{x^2}{4D_T t}\right) \exp[-\mu(x - x_n)] dx \qquad (1.8)$$

Equation (1.8) can be solved to give

$$\mu I_n - \frac{dI_n}{dx_n} = k \frac{S}{\sqrt{\pi D_T t}} \exp\left(-\frac{x_n^2}{4D_T t}\right)$$

or

$$\mu I_n - \frac{dI_n}{dx_n} = kc(x_n) \qquad (1.9)$$

[2] R. E. Pawel and T. S. Lundy, J. Appl. Phys., 34:1001 (1964); R. E. Pawel, Rev. Sci. Instr., 35:1066 (1964).

This method is of particular importance in two limiting cases.

(a) Weakly absorbed radiation — strong $\gamma$-rays. In this case $\mu I_n \ll dI_n/dx_n$ so that

$$-\frac{dI_n}{dx_n} = k \frac{S}{\sqrt{\pi D_T t}} \exp\left(-\frac{x_n^2}{4D_T t}\right)$$

and

$$\log_e\left(-\frac{dI_n}{dx_n}\right) = \text{constant} - \frac{x_n^2}{4D_T t} \tag{1.10}$$

Thus a plot of $\log_e(-dI_n/dx_n)$ versus $x_n^2$ should be a straight line of slope $-1/4D_T t$, which, the anneal time t being known, gives a value of the diffusion coefficient $D_T$ at T°K.

(b) Strongly absorbed radiation — weak $\beta$-rays. Here $\mu I_n \ll dI_n/dx_n$ so that

$$\mu I_n = k \frac{S}{\sqrt{\pi D_T t}} \exp\left(-\frac{x_n^2}{4D_T t}\right)$$

and

$$\log_e I_n = \text{constant} - \frac{x_n^2}{4D_T t} \tag{1.11}$$

A plot of $\log_e I_n$ versus $x_n^2$ should also be a straight line of slope $-1/4D_T t$, from which $D_T$ is calculated as before.

In the intermediate case of an isotope emitting both $\gamma$-rays and $\beta$-rays the difference in the count rate of the specimen with and without an absorber (i.e., the count rate of the $\gamma$-rays alone) is used as $I_n$ in (b) above.

The residual activity method of analysis is regarded as less reliable than the standard serial sectioning method, but it does allow us to measure diffusion coefficients several orders of magnitude smaller ($10^{-13}$ to $10^{-14}$ cm²/sec).

## 4.3. Surface Decrease Method

In this method the total activity of the specimen is measured as a function of time. No sectioning of the sample is necessary. As before, consider an element of thickness dx lying between x and x + dx. If the absorption is exponential, this element will contribute dI to the total activity I, so that

$$dI = \frac{kS}{\sqrt{\pi Dt}} \exp \left( -\frac{x^2}{4Dt} \right) \exp (-\mu x) dx$$

Integrating gives

$$I = \frac{kS}{\sqrt{\pi Dt}} \int_0^\infty \exp \left[ -\left( \frac{x^2}{4Dt} + \mu x \right) \right] \qquad (1.12)$$

which can be integrated to give

$$I = kS \exp(\mu^2 Dt) [1 - \operatorname{erf} \mu\sqrt{Dt}]$$

or

$$\frac{I}{I_0} = \exp(\mu^2 Dt) [1 - \operatorname{erf} \mu\sqrt{Dt}] \qquad (1.13)$$

The total activity I is measured as a function of time t. A master curve may be drawn giving $I/I_0$ as a function of $\mu^2 Dt$. The experimental curve is compared with a master to obtain a value of $\mu^2 Dt$, which, the value of the absorption coefficient $\mu$ and the anneal time t being known, yields the diffusion coefficient D.

This method can be useful for measuring small diffusion co-efficients ($10^{-13}$ to $10^{-15}$ cm$^2$/sec), particularly in the case of strong-ly absorbed radiation. However, one limitation is that an accurate value of $\mu$ is required and this is often difficult to ascertain in practice. The surface decrease method is not useful in the case of weakly absorbed (small $\mu$) radiation. The method is considerably less reliable than the serial sectioning method of analysis.

## 4.4. Autoradiography

All of the previously described methods of analysis have re-quired detection of the radiation by its ionizing action in some form of detector. There is one other method of detection, that of the action on photographic emulsions. The principle of the method is to determine the photographic density of the blackening of an exposed emulsion as a function of the distance from the interface of the specimen with the initial radioactive deposit.

The specimen is cut at a measured angle $\alpha$ (usually 90°) to the initial face, and the cut face is placed in contact with a piece of appropriate X-ray film. The type of film to be used and the length of exposure time are best found by experimentation. As a guide for normal diffusion samples with a few microcuries of initial activity, an exposure time of 24 hours is typical for the standard

types of X-ray film such as Kodak C or M. The blackening of the
emulsion (which is usually found to be directly proportional to the
concentration of activity) is measured with a microdensitometer.
To eliminate the uncertainty of determining the position of the
original interface, two specimens can be used with their active
faces welded together. The blackening of the emulsion is then sym-
metrical about the interface position.

A graph of photographic density versus distance is truly Gaus-
sian only in the case of a pure $\beta$-emitting isotope. In the case of
radioisotopes which are $\gamma$ emitters as well as $\beta$ emitters, a background blackening due to incompletely absorbed $\gamma$-rays from
atoms within the sample will be present. Various functions, such
as log (photographic density) versus distance, or log (photographic
density) versus distance squared, or photographic density versus
distance squared, give linear background plots. This "$\gamma$-tail" can
then be extrapolated back into the region of the diffusion zone per-
mitting a true log (photographic density) versus distance squared
graph to be drawn from the slope of which the diffusion coefficient
is easily calculated. The appropriate concentration distribution
equation is

$$c(x) = c(0)\exp\left(-\frac{x^2}{4Dt}\right)$$

and as the photographic density P is usually found to be directly
proportional to the concentration of active material c, we have

$$P(x) = P(0)\exp\left(-\frac{x^2}{4Dt}\right) \tag{1.14}$$

Hence a plot of $\log_e P(x)$ versus $x^2$ has a slope of $-1/4Dt$, from
which D may be calculated if the anneal time t is known.

The autoradiographic method is most suitable for pure $\beta$-
emitting isotopes of low or medium energies (less than 1 MeV).
The background correction is somewhat arbitrary, so that in general the accuracy of this method of analysis is not as high as that
of the standard serial sectioning method. However, it is very useful for low-energy $\beta$-emitting isotopes and for materials which
for various reasons may be difficult to section. The range of values
of the diffusion coefficients that can be measured by the autoradiographic method depends on the time of the anneal, the resolution

of the microdensitometer, and the type of film used. A typical
lower limit is $10^{-11}$ to $10^{-12}$ cm$^2$/sec.

The choice of the method of analysis depends on the material
used, the nature of the emitted particles from the radioactive
isotope, and the range of the diffusion coefficients being studied.

(a) The serial sectioning method is undoubtedly
the best although it is not practicable for weak $\beta$
emitters. It is the most accurate method for $\gamma$ emitting isotopes.

(b) The residual activity method can be used for both
$\beta$ - and $\gamma$ -emitting isotopes and can measure smaller
values of the diffusion coefficient. However, a value
of the absorption coefficient is required except in the
case of pure $\beta$ emitters.

(c) The surface decrease method suffers from the same
disadvantage as the residual activity method in that it
requires an accurate value of the absorption coefficient, as well as other problems from evaporation of
the isotope, etc.

(d) Autoradiography is most suitable for pure $\beta$ emitters
and can be used with isotopes emitting $\gamma$-rays as well
as $\beta$-rays but cannot be used with strong $\gamma$ emitters.

The choice of the form of analysis is summarized in Table I.

## 5.   THE TEMPERATURE DEPENDENCE
## OF THE DIFFUSION COEFFICIENT

Experimentally, the temperature dependence of the diffusion
coefficient can usually be expressed by an A r r h e n i u s   e q u a -
t i o n   of the form

$$D_T = D_0 \exp\left(-\frac{Q}{RT}\right)$$

where $D_T$ is the diffusion coefficient in cm$^2$/sec at a temperature
T°K, R is the universal gas constant (1.98 cal/mole $\cdot$ deg), Q is the
activation energy in cal/mole, and the constant $D_0$ is normally referred to as the frequency factor (cm$^2$/sec).

For self-diffusion in metals the values of $D_0$ usually lie between 0.1 and 10 cm$^2$/sec, often close to unity. If diffusion takes

## Table I

| Strong β and/or strong γ | Weak β | Weak γ | Diffusion coefficient (cm²/sec) | | Typical isotopes |
|---|---|---|---|---|---|
| | | | $10^{-5}-$ $10^{-11}$ | $10^{-10}-$ $10^{-14}$ | |
| √ | – | – | 1 | 1a 2 | $P^{32}$, $Lu^{170}$ $Hf^{171}$, $Ta^{177}$ |
| √ | √ | – | 1 | 1a | $Co^{60}$, $Mo^{99}$, $W^{185}$, $V^{48}$, $Cu^{64}$, $Ag^{110}$ |
| √ | √ | √ | 1 | 1a | $Mo^{99}$, $Au^{198}$ |
| – | √ | – | 2 4 3 | 2 3 | $Ni^{63}$, $Fe^{55}$, $In^{114}$, $Pd^{107}$ |
| – | – | √ | 1 | 1a | $Sn^{113}$, $U^{238}$ |
| – | √ | √ | 1 2 4 3 | 1a 2 3 | $Nb^{95}$, $Cr^{51}$, $Mn^{54}$, $Zr^{95}$ |
| √ | – | √ | 1 | 1a | $Rb^{83}$, $Zr^{97}$ |

1 —Serial sectioning (lathe).
1a —Serial sectioning (grinding or stripping).
2 —Residual activity.
3 —Surface decrease.
4 —Autoradiography.

place by more than one mechanism, or is affected by chemical gradients, then the measured $D_0$ values could be very different. Values of $D_0$ in pure grain boundary or dislocation diffusion also appear to be of the order of magnitude of unity, but the activation energies are usually less than for pure volume diffusion, often one-half the value.

## 6. MECHANISMS OF DIFFUSION

The energy barrier which opposes motion of an atom in a solid lattice is greater than that in a liquid or gas. Thus the activation energy Q for volume diffusion (diffusion through the lattice) is greater than that for liquids and gases. In addition to normal lattice diffusion, diffusion can also take place along grain boundaries, dislocations, or cracks in the material. The activation energies for grain boundary diffusion, dislocation diffusion, and surface diffusion along a crack are all less than that for volume diffusion through the lattice. This is so because grain boundaries, dislocations, and cracks are regions of higher energy than the lattice. The activation energies of diffusion in these three cases are related by

$$Q_{volume} > Q_{grain \ boundary} > Q_{surface}$$

The volume, grain boundary, and surface activation energies have been determined in a few cases and their relative values are of the order

$$Q_{volume} : Q_{grain \ boundary} : Q_{surface} = 4 : 2 : 1$$

Similarly, typical values of the frequency factor $D_0$ are related by

$$D_{0_{volume}} > D_{0_{grain \ boundary}} > D_{0_{surface}}$$

and the actual diffusion coefficients are related by

$$D_{surface} > D_{grain \ boundary} > D_{volume}$$

These three types of diffusion are illustrated in Fig. 2.

Several atomic mechanisms have been proposed for diffusion in solids. Each is illustrated in Fig. 3 and is briefly described below.

### 6.1. Vacancy Mechanism

The vacancy mechanism is simply the movement of an atom originally in a normal lattice site into a neighboring vacant site. In all materials, vacant sites or vacancies are present at all temperatures above absolute zero in equilibrium concentrations. Complex aggregates of vacancies such as divacancies and trivacancies are thought to exist. Normally diffusion by single

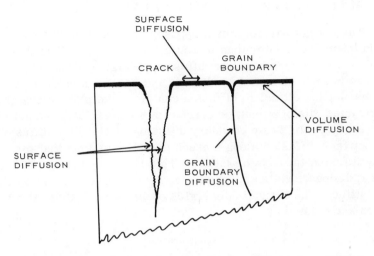

Fig. 2. Volume, grain boundary, and surface diffusion in a solid.

vacancies predominates over diffusion by divacancies, but divacancy diffusion may be appreciable at high temperatures. Enhanced diffusion in some systems at high temperatures has been
explained on the basis of a divacancy contribution.

## 6.2. Interchange or Exchange Mechanism

This is the simplest mechanism by which an atom can move.
It is simply the direct exchange of two nearest neighbor atoms.
This mechanism is unlikely in tightly bound crystals.

## 6.3. Ring Mechanism

The ring mechanism is merely a more general form of exchange mechanism consisting of a number (three or more) of atoms
forming a closed ring. Atomic diffusion then takes place by rota-
tion of the ring.

## 6.4. Interstitial Mechanism

The two preceding mechanisms can operate in a perfect crystal. Small atoms such as carbon will dissolve in metallic lattices
as impurities so as to occupy interstitial positions between the
solvent atoms. This has been shown to be the case for carbon in
$\alpha$-iron, for example. (Larger impurity atoms must occupy the

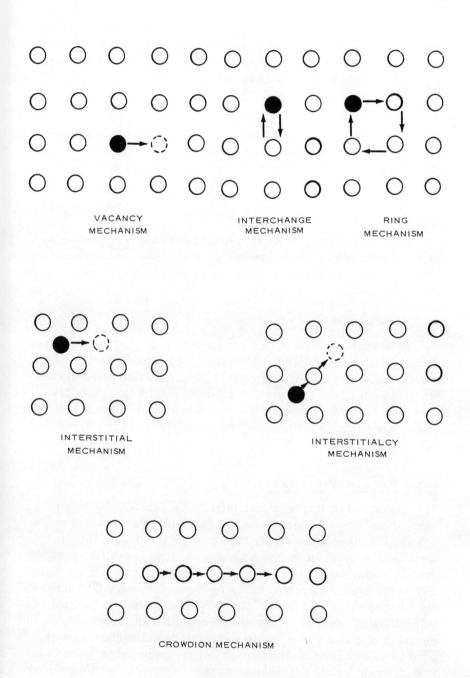

Fig. 3. Mechanisms of diffusion.

same atomic position as the solvent atoms when in solid solution; such positions are referred to as substitutional positions.) In the interstitial mechanism, diffusion takes place by atomic transport from one interstitial position to another.

## 6.5.  Interstitialcy Mechanism

In the interstitialcy mechanism, an interstitial atom moves from an interstitial position into an adjacent normal lattice site. In doing so it displaces the atom that had occupied that site to a new interstitial position.

## 6.6.  Crowdion Mechanism

A crowdion mechanism is a line imperfection consisting of n nearest neighbor atoms compressed into a space normally occupied by (n − 1) atoms. Diffusion then takes place by movement along the line of atoms.

## 6.7.  Dislocations

In practice, all materials are known to have intragranular defects called dislocations. Dislocations can provide easier paths for diffusion than a perfect lattice so that diffusion along dislocations is faster than true volume diffusion. Diffusion down dislocations has been proposed to explain enhanced diffusion rates at low temperatures, particularly, for example, in single crystals of the noble metals.

## 6.8.  Grain Boundary

The areas of intergranular misfit in a crystal also provide easy paths for a diffusing atom. In general diffusion along grain boundaries takes place faster than through the bulk of the material and in most cases the activation energy for diffusion along grain boundaries has been found to be smaller than that for lattice diffusion. Dislocation pipe diffusion, grain boundary diffusion, and surface diffusion all involve intricate paths within the crystal. Detailed kinetic analysis is therefore very difficult, especially as the numbers of dislocations, grain boundaries, and surfaces all depend to some extent on the temperature. These three types of diffusion, grain boundary, dislocation pipe, and surface diffusion, are relatively important only at low temperatures (less than about two-thirds of the melting point of the material).

## 7. SUMMARY OF TERMS

| | |
|---|---|
| Activation energy | The term Q in the Arrhenius equation $D = D_0 \exp(-Q/RT)$ |
| Autoradiography | Method of analysis utilizing the blackening of photographic film by radioactivity |
| Crowdion | Line imperfection consisting of n atoms occupying the space of $(n - 1)$ atoms in the regular lattice |
| Diffusion | Movement of matter through matter |
| Diffusion coefficient D | Proportionality constant in Fick's laws |
| Dislocation | Intragranular defect in the form of a line discontinuity |
| Fick's first law | Flux of diffusing species is proportional to concentration gradient: $J = -Ddc/dx$ |
| Fick's second law | Rate of change of concentration is proportional to second derivative of concentration with distance: $dc/dt = Dd^2c/dx^2$ |
| Frequency factor $D_0$ | The constant term $D_0$ in the Arrhenius equation $D = D_0 \times \exp(-Q/RT)$ |
| Grain boundary diffusion | Diffusion in the intergranular region |
| Interchange | Direct exchange of two nearest neighbor atoms |
| Interstitial mechanism | Movement from one interstitial position to another |

| | |
|---|---|
| Interstitialcy mechanism | Movement from an interstitial position to a normal lattice position |
| Residual activity | Sectioning method of analysis in which the activity remaining in the sample is measured |
| Self-diffusion | Diffusion of one substance into the same substance in the absence of a chemical gradient |
| Serial sectioning | Sectioning method of analysis in which the activity of each section is measured |
| Surface decrease | Method of analysis in which the total activity of the surface of the sample is measured as a function of time |
| Surface diffusion | Diffusion along the surface of a material or down a crack in it |
| Vacancy | Unoccupied lattice site in a crystal |
| Volume diffusion | Diffusion within the grains of a crystal |

## 8. SUGGESTED FURTHER READING

1.   Jost, W., Diffusion in Solids, Liquids, and Gases, Academic Press, New York (1952).
2.   Lazarus, D., Diffusion in Metals, in: Solid State Physics, Volume 10, H. Ehrenreich, F. Seitz, and D. Turnbull, eds., Academic Press, New York (1960).
3.   Shewmon, P. G., Diffusion in Solids, McGraw-Hill, New York (1963).
4.   Leymonie, C., Radioactive Tracers in Physical Metallurgy, Chapman and Hall, London (1963).
5.   Manning, J. R., Diffusion Kinetics for Atoms in Crystals, Van Nostrand, Princeton (1968).
6.   Girifalco, L. A., Atomic Migration in Crystals, Blaisdell Publishing Co., New York (1964).

7.  Diffusion in Body-Centered Cubic Metals, American Society
    for Metals, Metals Park, Ohio (1965).

*Chapter 2*

# Empirical and Semi-Empirical Diffusion Relations

Many empirical relations between the diffusion parameters Q and $D_0$ and other physical properties of materials have been proposed during the past 30 to 40 years. In general the relations are of two types:

    (a)  those derived without any consideration of diffusion mechanism, and

    (b)  those derived on the basis of some assumed mechanism.

The purpose here is to present the more important relations and to discuss their usefulness in estimating the diffusion parameters $D_0$ and Q for systems for which experimental data are not yet available.

## 1. THE DUSHMAN AND LANGMUIR RELATION

In 1922 Dushman and Langmuir[1] proposed a relation for diffusion in metallic systems. They derived the relation

$$D = \frac{Qa^2}{Nh} \exp\left(-Q/RT\right)$$

where $a$ is the lattice constant, N is Avogadro's number, and h is

---

[1] S. Dushman and I. Langmuir, Phys. Rev., 20:113 (1922).

Planck's constant. Thus,

$$D_0 = \frac{Qa^2}{Nh} = 1.04 \times 10^{-3} \, Qa^2 \qquad (2.1)$$

where $a$ is measured in Å ($1 \times 10^{-8}$ cm) and Q is measured in kcal/mole. The values of $D_0$ vary over the fairly narrow limits of 0.2 to 0.9 cm$^2$/sec for the various elements. However, the equation is useful since it permits values of $D_0$ and Q to be calculated from one experimental measurement of the diffusion coefficient at one temperature. If we assume a typical value of $D_0$ of 0.5 cm$^2$/sec, this is essentially the same as calculating Q from

$$Q = -RT \ln 2D \qquad (2.2)$$

where Q is the activation energy in cal/mole, R is the gas constant (1.98 cal/mole · deg), T is the temperature in degrees Kelvin, and D is the diffusion coefficient in cm$^2$/sec.

Equations (2.1) and (2.2) are in good agreement with self-diffusion data in pure metals (except for the "anomalous" b.c.c. metals) and for much of the impurity and alloy data.

## 2. SIMPLE EMPIRICAL CORRELATIONS BETWEEN THE SELF-DIFFUSION ACTIVATION ENERGY AND VARIOUS PHYSICAL PROPERTIES OF METALS

A number of empirical correlations have been suggested between the self-diffusion activation energy Q and various physical properties of materials such as melting point, heats of fusion and vaporization, and elastic moduli since all of these and atomic diffusion in solids depend on the binding forces between atoms. In such correlations the self-diffusion activation energy Q is related to the physical property P by a relation of the form

$$Q = AP \qquad \text{or} \qquad Q = B/P \qquad (2.3)$$

where A and B are proportionality constants. Some of these correlations are very useful in predicting values of the self-diffusion activation energy with a fair degree of accuracy. However, care must be taken in attempting to give the correlation any type of fundamental interpretation.

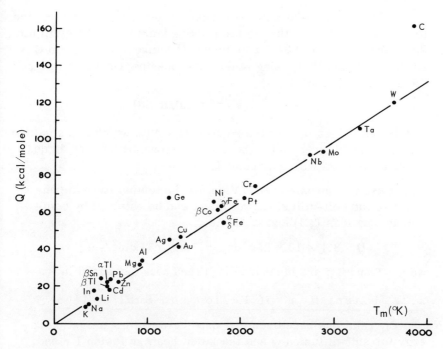

Fig. 4. Self-diffusion activation energy as a function of melting point.

## 2.1.   Melting Point

The melting point, $T_m$, is one physical property of materials that is closely related to atomic transport. This has led to one of the most useful correlations

$$Q = AT_m$$

where A is a constant. Values of the self-diffusion activation energy are plotted as a function of the melting point in Fig. 4. An average value of the constant A for all structures is 33.7 cal/mole· deg, giving

$$Q = 33.7T_m \qquad (2.4)$$

However, a value of A = 38.0 cal/mole · deg gives a better fit for f.c.c. structures and a value of A = 32.5 cal/mole · deg a better fit for b.c.c. structures. The elements Sn, Ge, C, $\gamma$-U, $\beta$-Ti, and $\beta$-Zr (all elements in Group IV of the periodic table) do not fit this simple correlation at all well.

Oshyarin[2] derived a relation between the atomic radius r, the force constant K, and the atomic packing fraction $f$ (0.78 for f.c.c., 0.68 for b.c.c., and 0.34 for diamond structures) of the form $Q = Kr^3/f^{2/3}$. From this, using other relationships, Oshyarin deduced the simple relation

$$Q = \frac{\text{constant } T_m}{f^{2/3}} [\text{cal/mole}]$$

This gives values of A in the relation $Q = AT_m$ which are smallest for b.c.c. structures and highest for diamond structures, in qualitative agreement with experiment.

A useful approximate relation for the temperature dependence of the self-diffusion coefficient can be obtained by combining equations (2.1) and (2.4), i.e.,

$$D = 3.4 \times 10^{-5} T_m a^2 \exp(-17.0 T_m/T) [\text{cm}^2/\text{sec}]$$

where T and $T_m$ are measured in °K and $a$ is measured in Å.

## 2.2. Latent Heat of Fusion and Sublimation

Correlations have been proposed between the activation energy for self-diffusion Q and the latent heats of fusion $L_f$ and sublimation $L_s$. Approximate relationships are $Q = 16.5L_f$ and $Q = 0.65L_s$, where the latent heats are given in kcal/mole. Such correlations are much less consistent than the melting point correlation, and are not generally used.

## 2.3. Compressibility

Compressibility is another physical property of materials that depends on the binding energy between atoms. Such a correlation of the form $Q = B/\chi$, where $\chi$ is the compressibility, has been given by Gibbs.[3]

## 2.4. Coefficient of Linear Expansion

Another physical property directly related to compressibility is the coefficient of linear expansion. An analysis by the author[4] shows that an approximate correlation exists between the self-diffusion activation energy Q and the room temperature coefficient of linear expansion, $\alpha$:

$$Q = \frac{700}{\alpha} [\text{kcal/mole}]$$

[2]B. N. Oshyarin, Phys. Status Solidi, 3:K61 (1963).
[3]G. B. Gibbs, C.E.G.B. Report RD/B/N.355, Nov. 1964.
[4]G. Askill, Phys. Status Solidi, 11:K49 (1965).

where $\alpha$ is measured in ppm/°C, gives a good fit to the majority of the experimental data with an accuracy of about 10%.

## 3. RELATIONS INVOLVING ELASTIC MODULI

A relation of the form Q = A (elastic modulus) has been given by Buffington[5] using the vacancy model of diffusion. The relation is $Q = KE_0 a^3$, where $E_0$ is an appropriate elastic constant, $a$ is the lattice parameter, and K is a proportionality constant depending only on the crystal structure.

A relation between the frequency factor $D_0$ and the activation energy Q was proposed by Zener[6] using elasticity theory. Using random walk theory he deduced the relation

$$D_0 = a^2 \nu \exp (\lambda \beta Q / RT_m)$$

or

$$\ln \left( \frac{D_0}{a\nu} \right) = \frac{Q}{T_m} \frac{\lambda \beta}{R}$$

where $\lambda$ is the fraction of energy which goes into straining the lattice and $\beta$ is the dimensionless constant $-d(\mu/\mu_0)/d(T/T_m)$ and $\mu$ is an appropriate elastic constant. Values of $\beta$ lie in the range 0.25 to 0.45 for most metals, and since $\lambda = 0.6$ for f.c.c. and 0.8 for b.c.c. metals, the values of $\lambda \beta$ are between 0.15 and 0.35.

A relation of the form $D_0 = a^2 \nu K \exp (Q/RT_m)$ was found empirically by Diennes with $K = 10^{-6}$. The value of K together with $Q/T_m = 34$ gives a value of $\lambda \beta = 0.2$. A more extensive analysis has been made by LeClaire[7] considering mechanisms other than the vacancy mechanism.

## 4. AN EXTENDED MELTING POINT CORRELATION

The self–diffusion activation energy has been shown to be dependent on the melting point of a metal. However, one would also expect the self–diffusion activation energy to be dependent on the valence and crystal structure of the material. Sherby and

[5]F. S. Buffington and M. Cohen, Acta Met., 2:660 (1954).
[6]C. Zener, Acta Cryst., 3:346 (1950).
[7]A. D. LeClaire, Acta Met., 1:438 (1953).

Simnad[8] have given an empirical analysis of the variation of the self-diffusion activation energy Q with the melting point $T_m$, valence V, and crystal structure factor $K_0$. From a plot of log D versus $T/T_m$ it is evident that the majority of the diffusion data falls into three distinct groups: (a) b.c.c., (b) f.c.c. and c.p.h., and (c) metals of diamond structure, giving the relation $D = D_0 \exp(-KT_m/T)$, where the constant K varies with the valence V according to $K = K_0 + V$ and $K_0$ is 14 for b.c.c. metals, 17 for f.c.c. and c.p.h. metals, and 21 for metals of the diamond structure.

Thus $Q = (K_0 + V)RT_m$ gives a much greater fit to all the experimental data than the simpler relation $Q = $ constant $T_m$. The effective valence of the transition metals can be deduced from their experimental values of the self-diffusion activation energies. They seem to be fairly constant for the elements in each group of the periodic table. Values of V are 1.5 for Group IVB (Ti, Zr, Hf), 3.0 for Group VB (V, Nb, Ta), 2.8 for Group VIB (Cr, Mo, W), 2.6 for Group VIIB (Mn, Re), and 2.5 for other transition elements. The relation $Q = (K_0 + V)RT_m$ appears to be the best fit to all the self-diffusion data, and therefore is the best relation for predicting unknown self-diffusion activation energies.

A similar relation has been deduced by LeClaire.[9] He found that the relation $Q = RT_m (K + 1.5V)$ was a slightly better fit with K = 13 for b.c.c. metals, 15.5 for f.c.c. and c.p.h. metals, and 20 for metals of the diamond structure.

## 5. EMPIRICAL RELATIONS FOR IMPURITY DIFFUSION AND SELF-DIFFUSION IN ALLOYS

Impurity diffusion has been studied in detail in several solvent elements. However, no single correlation between the diffusion parameters $D_0$ and Q and properties of the impurity elements such as valence and size factor has been found to fit more than one or two elements. Extensive analyses of impurity diffusion in the noble metals have been made by Lazarus[10] and LeClaire,[11] who

---

[8]O. Sherby and M. T. Simnad, Trans. ASM, 54:227 (1961).
[9]A. D. LeClaire, Diffusion in Body-Centered Cubic Metals, American Society for Metals, Metals Park, Ohio (1965), page 10.
[10]D. Lazarus, Phys. Rev., 93:973 (1954).
[11]A. D. LeClaire, Phil. Mag., 7(73):141 (1962).

showed that the activation energies for impurity solutes decrease
with increasing valence. Experimental results are in excellent
agreement.

Impurity diffusion in the transition elements follows a different pattern. In molybdenum, for example, the $D_0$ and Q values
for impurity diffusion of $Co^{60}$, $Nb^{95}$, and $W^{185}$ are all greater than
self-diffusion although no simple correlation with size factor is
apparent. In contrast to this, the activation energies for impurity
diffusion in niobium are all less than that for self-diffusion. Titanium is the transition element for which the largest number of
impurity solutes have been studied. They are Sc, V, Cr, Mn, Fe,
Co, Ni, Nb, Mo, Sn, Ag, and P. An analysis of the low- and high-
temperature impurity diffusion parameters $D_0$ and Q gave the relations

$$D_{0\,(low)} = 2 \times 10^{-4} \exp{(0.29S_f)} \quad \text{and} \quad D_{0\,(high)} = 1 \exp{(0.20S_f)}$$

where $S_f$ is the size factor. These relations hold only for solutes
of negative size factor. No correlation was found for the impurity
activation energies. It seems, therefore, that no general empirical
relations have been found for the impurity diffusion parameters
$D_0$ and Q. General trends seem to be obeyed only for specific elements. From these one can estimate how the parameters $D_0$ and
Q may be expected to vary for impurities similar to those studied.

Several alloy systems of transition and nontransition elements
have been studied in detail. As for impurity diffusion, most relations that have been found hold true only for that particular system. However, one simple relation that has been found to hold
true for many dilute alloy systems is $Q = Q_p[1 + ac]$, where $Q_p$
is the impurity activation energy in a pure metal A, and $Q$ is the
impurity activation energy for an alloy of concentration c of B
in A. A similar relation often holds true for the frequency factor
$D_0$ of the form

$$\ln{D_0} = \ln{D_{0P}} \, [1 + ac]$$

where $a$ is a constant depending on the system.

## 6.  SUMMARY OF EMPIRICAL RELATIONS

The various empirical and semi-empirical relations that have
been discussed in this chapter are summarized here. They are
given in the order of their importance and usefulness.

Self-diffusion
activation energy

$$Q = (K_0 + V)RT_m$$

Self-diffusion coefficient

$$D = 3.4 \times 10^{-5} T_m a^2 \exp(-17T_m/T)$$

$$Q = 33.7T_m$$

Self-diffusion frequency
factor

$$Q = -RT \ln 2D$$

$$D_0 = a^2 \nu \exp(\lambda \beta Q / RT_m)$$

$$D_0 = 1.04 \times 10^{-3} Q a^2$$

$$Q = 700/\alpha$$

Alloy diffusion

$$Q = 16.5L_f$$

$$Q = Q_0[1 + \text{constant} \times \text{conc.}]$$

$$Q = 0.65L_s$$

$$\ln D_0 = \ln D_{0p}[1 + \text{constant} \times \text{conc.}]$$

where

$D$ = diffusion coefficient in $cm^2/sec$
$D_0$ = frequency factor in $cm^2/sec$
$a$ = interatomic distance or lattice constant in Å ($1 \times 10^{-8}$ cm)
$T_m$ = melting point in °K
$T$ = temperature in °K
$\nu$ = vibrational frequency in $sec^{-1}$
$R$ = gas constant (1.98 cal/mole · deg)
$L_f$ = latent heat of fusion in kcal/mole
$L_s$ = latent heat of sublimation in kcal/mole
$\lambda$ = constant (0.6 for b.c.c., 0.8 for f.c.c. metals)
$\beta$ = constant ($\approx 1/2$)
$\alpha$ = coefficient of linear expansion in ppm/°C
$K_0$ = constant (14 for b.c.c., 17 for f.c.c. and c.p.h., and 21 for diamond structure)
$V$ = valence

*Chapter 3*

# Radioactive Tracer Diffusion Data in Metals, Alloys, and Oxides

## 1. INTRODUCTION

The data given in these tables cover radioactive tracer diffusion data of metals in metals, alloys, and simple oxides published in the literature from 1938 to December 1968. It is divided into four parts.

I.   Self-diffusion in pure metals.
II.  Impurity diffusion in pure metals.
III. Self- and impurity diffusion in metal alloys.
IV.  Self- and impurity diffusion in simple metal oxides.

The data are presented as the constants of the Arrhenius equation $D = D_0 \exp(-Q/RT)$, when D is the diffusion coefficient in $cm^2/sec$ at absolute temperature T, Q is the activation energy in kcal/mole, and R is the gas constant (1.98 cal/mole · deg).

The following abbreviations are used throughout:

| | | | |
|---|---|---|---|
| S.S. | = serial sectioning | P | = polycrystal |
| R.A. | = residual activity | S | = single crystal |
| S.D. | = surface decrease | ⊥c | = perpendicular to c direction |
| A.R.G. | = autoradiography | ‖ c | = parallel to c direction |
| at | = atomic | 99.95 | = 99.95% purity |
| wt | = weight | | |

In general, the most reliable measurements are the more recent ones using high-purity materials, serial sectioning analysis, and giving $D_0$ values in the range 0.1 to 10 $cm^2/sec$. Some sections, for example, self-diffusion in silver, have several entries. It is often difficult to determine which set or sets of data are most representative of the system. One has to consider the purity of the samples, method of analysis, number of data points, and the general experimental arrangement. Those sets of data which, in the opinion of the author based on the above criteria, are most representative of the particular system are indicated by asterisks: *** ***.

The four main methods of analysis of the diffusion samples are serial sectioning, residual activity, surface decrease, and autoradiography. These are discussed in some detail in Chapter 1.

## 2.  SOME PREDICTED DIFFUSION PARAMETERS

In the event that a certain set of experimental data is not available, the various empirical relations that are summarized at the end of Chapter 2 can be used to estimate the diffusion parameters. The following are estimates of the self-diffusion activation energies Q for some elements for which few or no experimental data are available.

| Element | $T_m$ (°K) | Q (kcal/mole) |
| --- | --- | --- |
| δ -Mn | 1533 | 51 |
| γ -Mn | 1533 | 59 |
| Ru | 2750 | 107 |
| Rh | 2240 | 87 |
| Re | 3440 | 135 |
| Os | 2975 | 116 |
| Ir | 2720 | 107 |

Diffusion experiments in the high-temperature b.c.c. phases of the Group IVB metals (Ti, Zr, and U) have shown that the Arrhenius plots for each of these elements and for each of the many impurity elements that have been studied are nonlinear. The activation energies are about 1/2 that expected from the various empirical relations, and the frequency factors are all typically $10^{-3}$ $cm^2/sec$. Diffusion in these metals has been called anomalous, and it is predicted that diffusion in the other Group IV elements, Hf, Th, and also Pa, Np, and Pu, will also be anomalous. On this basis the following are estimates of the self-diffusion activation energy Q in these materials. The frequency factors are all expected to be about $10^{-3}$ cm$^2$/sec.

| Element | $T_m$ (°K) | Q (kcal/mole) |
|---|---|---|
| $\beta$ -Hf | 2520 | 34 |
| $\beta$ -Th | 2120 | 32 |
| b.c.c. Pa | 1200 | 19 |
| f.c.c. Pa | 1200 | 22 |
| b.c.c. Np | 810 | 13 |
| f.c.c. Np | 810 | 15 |
| b.c.c. Pu | 810 | 13 |
| f.c.c. Pu | 810 | 15 |

# Part I

## Self-Diffusion in Pure Metals

| Material | Temperature range (°C) | Form of analysis | Activation energy Q (kcal/g-atom) | Frequency factor, $D_0$ (cm²/sec) | Reference No. | Year |
|---|---|---|---|---|---|---|
| ALUMINUM | Tracer ²⁶Al | | | | | |
| | | | | | | |
| P 99.99 | 450-650 | S.S. | 34.0 | 1.71 | 1 | (1962) |
| S | 400-610 | R.A. | 34.5 | — | 2 | (1968) |
| | | | | | | |
| ANTIMONY | Tracer ¹²⁴Sb | | | | | |
| | | | | | | |
| S⊥c 99.998 | 500-620 | S.S. | 44.4 | 16.6 | 3 | (1964) |
| S∥c 99.998 | 500-620 | S.S. | 47.1 | 22.1 | 3 | (1964) |
| P 99.99 | 473-583 | R.A. | 39.5 | 1.05 | 4 | (1965) |
| S∥c 99.999 | 500-630 | — | 48.0 | 56.0 | 5 | (1965) |
| S⊥c 99.9999 | 500-630 | — | 35.8 | 0.1 | 5 | (1965) |
| | | | | | | |
| BERYLLIUM | Tracer ⁷Be | | | | | |
| | | | | | | |
| S⊥c 99.75 | 565-1065 | R.A. | 37.6 | 0.52 | 6 | (1965) |
| S∥c 99.75 | 565-1065 | R.A. | 39.4 | 0.62 | 6 | (1965) |
| P 99.9 | 650-1200 | R.A. | 38.4 | 0.36 | 7 | (1968) |
| | | | | | | |
| CADMIUM | Tracer ¹¹⁵Cd | | | | | |
| | | | | | | |
| S⊥c 99.5 | 130-280 | S.S. | 19.1 | 0.10 | 8 | (1955) |
| S∥c 99.5 | 130-280 | S.S. | 18.2 | 0.05 | 8 | (1955) |
| S,P | 200-285 | S.D. | 19.7 | 0.14 | 9 | (1958) |
| S | 180-300 | — | 20.6 | 0.68 | 10 | (1967) |

31

| Material | Temperature range (°C) | Form of analysis | Activation energy Q (kcal/g-atom) | Frequency factor, $D_0$ (cm²/sec) | Reference No. | Year |
|---|---|---|---|---|---|---|

**CARBON (Graphite)   Tracer $^{14}$C**

| Natural crystals | 2000-2200 | — | 163 | 0.4-14.4 | 11 | (1957) |

**CHROMIUM        Tracer $^{51}$Cr**

| P 99.94 | 1000-1350 | A.R.G. | 76.0 | | 12 | (1957) |
| P | 1000-1350 | — | 85.0 | 45.0 | 13 | (1959) |
| P 99.8 | 950-1260 | S.D.,R.A. | 52.7 | $1.5 \times 10^{-4}$ | 14 | (1959) |
| P 99.95 | 700-1350 | A.R.G. | 76.0 | 0.4 | 15 | (1959) |
| P 99.87 | 1080-1340 | A.R.G. | 62.4 | $1.65 \times 10^{-3}$ | 16 | (1960) |
| P 99.96 | 1060-1400 | R.A. | 59.2 | $6.47 \times 10^{-2}$ | 17 | (1962) |
| P 99.99 | 1200-1600 | R.A. *** | 73.2*** | 0.28 | 18 | (1962) |
| P 99.98 | 1030-1545 | S.S. *** | 73.7*** | 0.2 | 19 | (1965) |

**β-COBALT        Tracer $^{60}$Co**

| P 98.7 | 1050-1250 | S.D. | 67.0 | 0.367 | 20 | (1951) |
| P 99.9 | 1000-1250 | S.D. | 61.9 | 0.032 | 21 | (1951) |
| P 98.4 | 1000-1300 | R.A. | 62.0 | 0.2 | 22 | (1952) |
| P 99.9 | 1100-1405 | S.S. *** | 67.7*** | 0.83 | 23 | (1955) |
| P | 772-1048 | | 65.4 | 0.50 | 24 | (1962) |
| P | 1192-1297 | | 62.2 | 0.17 | 24 | (1962) |
| P 99.5 | 1047-1311 | R.A. | 68.7 | 1.66 | 25 | (1965) |

**COPPER        Tracer $^{64}$Cu**

| P | 880-1030 | S.S. | 61.4 | 47.0 | 26 | (1939) |
| P | 750-850 | | 57.2 | 11.0 | 27 | (1939) |
| P | 860-970 | S.S. | 45.1 | 0.10 | 28 | (1942) |
| S | 860-970 | S.S. | 49.0 | 0.6 | 28 | (1942) |
| P | 650-850 | S.D. | 46.8 | 0.3 | 29 | (1942) |
| P | 685-1060 | S.S.*** | 47.12*** | 0.2 | 30 | (1954) |
| P | 850-1050 | S.S.*** | 49.56*** | 0.621 | 31 | (1955) |
| P 99.99 | 863-1057 | S.S. | 48.2 | 0.33 | 32 | (1964) |

| Material | Temperature range (°C) | Form of analysis | Activation energy Q (kcal/g-atom) | Frequency factor, $D_0$ (cm²/sec) | Reference No. | Year |
|---|---|---|---|---|---|---|
| Copper | Tracer $^{64}$Cu | (continued) | | | | |
| | | | | | | |
| P | 780-890 | — | 49.2 | 2.32 | 33 | (1967) |
| S | 700-990 | — | 46.9 | — | 2 | (1968) |
| | | | | | | |
| GERMANIUM | Tracer $^{71}$Ge | | | | | |
| | | | | | | |
| S | 780-930 | S.S. | 73.5 | 87.0 | 34 | (1954) |
| S | 766-928 | S.S.*** | 68.5*** | 7.8 | 35 | (1956) |
| S | 730-916 | R.A. | 69.4 | 10.8 | 36 | (1961) |
| S | 730-916 | R.A. | 77.5 | 44.0 | 36 | (1961) |
| | | | | | | |
| GOLD | Tracer $^{198}$Au | | | | | |
| | | | | | | |
| P | 720-970 | R.A. | 51.0 | 2 | 37 | (1938) |
| P | 800-1000 | A.R.G. | 45.0 | — | 38 | (1952) |
| P | 775-1060 | A.R.G. | 45.3 | 0.265 | 39 | (1954) |
| P 99.96 | 730-1030 | A.R.G. | 45.3 | 0.26 | 40 | (1955) |
| S 99.999 | 600-954 | S.S. | 39.36 | 0.031 | 41 | (1956) |
| P 99.99 | 720-1000 | S.S. | 42.9 | 0.14 | 42 | (1957) |
| P 99.95 | 700-1050 | S.S.*** | 41.7*** | 0.091 | 43 | (1957) |
| P 99.93 | 700-900 | R.A.*** | 42.1*** | 0.117 | 44 | (1963) |
| S 99.97 | 850-1050 | S.S.*** | 42.26*** | 0.107 | 45 | (1965) |
| P 99.999 | 706-1010 | S.S. | 45.7 | 0.15 | 46 | (1965) |
| S | 600-900 | S.S. | 40.0 | 2 | (1968) | |
| | | | | | | |
| β-HAFNIUM | Tracer $^{181}$Hf | | | | | |
| | | | | | | |
| P 97.9 | 1795-1995 | S.S. | 38.7 | $1.2 \times 10^{-3}$ | 47 | (1965) |
| | | | | | | |
| INDIUM | | | | | | |
| | | | | | | |
| P,S 99.998 | 50-160 | S.S. | 17.9 | 1.02 | 48 | (1952) |
| S⊥c 99.99 | 44-144 | S.S. | 18.7 | 3.7 | 49 | (1959) |
| S∥c 99.99 | 44-144 | S.S. | 18.7 | 2.7 | 49 | (1959) |
| P 99.9999 | 130-149 | R.A. | 21.2 | 8.0 | 50 | (1965) |

| Material | Temperature range (°C) | Form of analysis | Activation energy Q (kcal/g-atom) | Frequency factor, $D_0$ (cm²/sec) | Reference No. | Year |
|---|---|---|---|---|---|---|
| α-IRON | Tracer $^{55}$Fe, $^{59}$Fe | | | | | |
| P | 715-890 | R.A. | 77.2 | 34,000 | 51 | (1948) |
| P | 700-900 | R.A. | 73.2 | 2300 | 52 | (1950) |
| P 99.97 | 800-900 | S.D. | 59.7 | 5.8 | 53 | (1951) |
| P | 800-900 | R.A. | 48.0 | 0.1 | 54 | (1952) |
| P | 700-900 | | 47.4 | 0.019 | 55 | (1955) |
| P 99.96 | 650-850 | S.D. | 67.1 | 530 | 56 | (1955) |
| P,S | 775-900 | | 64.1 | 18 | 57 | (1958) |
| P | 800-900 | R.A. | 67.24 | 118 | 58 | (1958) |
| P,S 99.97 | 700-790 | S.S.,R.A. *** | 60.0 *** | 2.0 | 59 | (1961) |
| P,S 99.97 | 790-900 | S.S.,R.A. *** | 57.2 *** | 1.9 | 59 | (1961) |
| P 99.998 | 860-900 | A.R.G. *** | 57.3 *** | 2.0 | 60 | (1963) |
| P 99.98 | 750-850 | | 72.17 | 900 | 61 | (1964) |
| P 99.95 | 638-768 | R.A. *** | 60.7 *** | 27.5 | 62 | (1966) |
| P 99.95 | 808-884 | R.A. *** | 57.5 *** | 2.01 | 62 | (1966) |
| γ-IRON | Tracer $^{55}$Fe, $^{59}$Fe | | | | | |
| P | 935-1110 | R.A. | 48.0 | $1.0 \times 10^{-3}$ | 51 | (1948) |
| P | 950-1400 | S.D. | 67.9 | 0.58 | 53 | (1951) |
| P | 1000-1360 | R.A. | 74.2 | 5.8 | 52 | (1950) |
| P | 1000-1300 | R.A. | 68.0 | 0.76 | 54 | (1952) |
| P 99.1 | 950-1250 | R.A. | 68.0 | 0.7 | 63 | (1953) |
| P | 900-1200 | | 67.9 | 1.3 | 55 | (1955) |
| P | 1000-1300 | S.S. | 67.0 | 0.44 | 64 | (1956) |
| P | 900-1200 | | 64.0 | 0.16 | 65 | (1957) |
| P | 1000-1250 | R.A. | 63.5 | 0.11 | 66 | (1958) |
| P 99.6 | 1000-1200 | A.R.G. | 64.0 | 0.16 | 15 | (1959) |
| P,S 99.97 | 1064-1395 | S.D.,R.A.*** | 64.5*** | 0.18 | 59 | (1961) |
| P | 950-1200 | S.S. | 67.7 | 2.5 | 67 | (1962) |
| P 99.998 | 1156-1350 | A.R.G.*** | 64.0 *** | 0.22 | 60 | (1963) |
| P 99.94 | 1075-1340 | R.A. | 67.8 | 1.05 | 68 | (1965) |
| δ-IRON | Tracer $^{55}$Fe, $^{59}$Fe | | | | | |
| P | 1405-1520 | R.A. | 42.4 | 0.019 | 69 | (1961) |
| P 99.96 | 1415-1510 | S.S.,R.A.*** | 57.0 *** | 1.9 | 70 | (1963) |
| P 99.998 | 1405-1515 | A.R.G. | 61.7 | 6.8 | 60 | (1963) |
| P 99.95 | 1428-1492 | S.S.*** | 57.5 *** | 2.01 | 62 | (1962) |

| Material | Temperature range (°C) | Form of analysis | Activation energy Q (kcal/g-atom) | Frequency factor, $D_0$ (cm²/sec) | Reference No. | Year |
|---|---|---|---|---|---|---|
| LEAD | Tracer $^{210}$Pb | | | | | |
| P | 260-320 | S.D. | 27.9 | 6.56 | 71 | (1932) |
| S,P | 190-270 | S.S. | 25.7 | 1.17 | 72 | (1954) |
| S 99.999 | 175-322 | S.S.*** | 24.21*** | 0.281 | 73 | (1955) |
| S 99.99 | 207-323 | S.S.*** | 26.06*** | 1.372 | 74 | (1961) |
| MAGNESIUM | Tracer $^{28}$Mg | | | | | |
| P 99.92 | 470-630 | S.S. | 32.0 | 1.0 | 75 | (1954) |
| S⊥c | 467-635 | S.S. | 32.5 | 1.5 | 76 | (1956) |
| S∥c | 467-635 | S.S. | 32.2 | 1.0 | 77 | (1956) |
| MOLYBDENUM | Tracer $^{99}$Mo | | | | | |
| P | 1800-2200 | R.A. | 115 | 4 | 77 | (1959) |
| P | 1800-2200 | R.A. | 114 | 4 | 13 | (1959) |
| P | 1700-1900 | | 111 | 2.77 | 78 | (1960) |
| P | 1600-2200 | | 100.8 | 0.38 | 79 | (1961) |
| S 99.99 | 1850-2350 | S.S.*** | 92.2*** | 0.1 | 80 | (1963) |
| P 99.98 | 1850-2350 | S.S.*** | 96.9*** | 0.5 | 80 | (1963) |
| P 99.97 | 2155-2540 | R.A.*** | 110.0*** | 1.8 | 81 | (1964) |
| NICKEL | Tracer $^{63}$Ni | | | | | |
| P 99.93 | 870-1250 | S.D.,R.A. | 66.8 | 1.27 | 82 | (1956) |
| P | 1100-1175 | S.S. | 63.8 | 0.4 | 83 | (1957) |
| S 99.95 | 700-1100 | A.R.G. | 65.9 | 0.48 | 84 | (1958) |
| P | 1150-1400 | S.S. | 69.8 | 3.36 | 85 | (1959) |
| P 99.92 | 1150-1400 | R.A. | 71.0 | 5.12 | 86 | (1959) |
| S | 680-830 | R.A. | 69.7 | 5.8 | 87 | (1961) |
| S 99.999 | 950-1020 | S.D. | 66.8 | 1.9 | 88 | (1961) |
| | 700-1100 | | 68.1 | 1.70 | 24 | (1962) |
| P 99.98 | 1085-1300 | S.S.*** | 69.5*** | 2.59 | 89 | (1963) |
| P | 950-1250 | R.A. | 67.0 | 2.4 | 90 | (1963) |
| P 99.95 | 1042-1404 | S.S.*** | 68.0*** | 1.9 | 32 | (1964) |
| P 99.97 | 1155-1373 | R.A. | 64.9 | 1.11 | 25 | (1965) |
| S,P 99.99 | 475-650 | S.D. | 66.8 | 1.9 | 91 | (1965) |
| S,P 99.99 | 675-750 | S.D. | 66.8 | 1.9 | 92 | (1965) |
| S 99.9 | 900-1200 | R.A.*** | 70.1*** | 2.59 | 93 | (1966) |

| Material | Temperature range (°C) | Form of analysis | Activation energy Q (kcal/g-atom) | Frequency factor, $D_0$ (cm²/sec) | Reference | |
|---|---|---|---|---|---|---|
| | | | | | No. | Year |

NICKEL        Tracer $^{63}$Ni (continued)

| | | | | | | |
|---|---|---|---|---|---|---|
| P 99.9 | 900-1200 | R.A.*** | 69.2*** | 2.22 | 93 | (1966) |
| S 99.999 | 980-1400 | R.A. | Nonlinear | | 94 | (1968) |
| P | 1000-1400 | R.A. | 60.5 | 9.96 | 95 | (1968) |

NIOBIUM        Tracer $^{95}$Nb

| | | | | | | |
|---|---|---|---|---|---|---|
| P 99.4 | 1535-2120 | S.S. | 105.0 | 12.4 | 96 | (1960) |
| P 99.8 | 1700-2100 | A.R.G. | 95.0 | 1.3 | 97 | (1962) |
| P,S 99.99 | 878-2395 | S.S.*** | 96.0*** | 1.1 | 98 | (1965) |
| S 99.95 | 1700-2100 | R.A. | 115.0 | 49.0 | 99 | (1964) |
| P 99.95 | 1700-2100 | R.A. | 110.0 | 17.0 | 99 | (1964) |

PALLADIUM        Tracer $^{103}$Pd

| | | | | | | |
|---|---|---|---|---|---|---|
| S 99.999 | 1060-1500 | S.S. | 63.6 | 0.205 | 100 | (1964) |

PHOSPHORUS        Tracer $^{32}$P

| | | | | | | |
|---|---|---|---|---|---|---|
| P | 0-44 | S.S. | 9.4 | $1.07 \times 10^{-3}$ | 101 | (1955) |

PLATINUM        Tracer $^{195}$Pt, $^{199}$Pt

| | | | | | | |
|---|---|---|---|---|---|---|
| P 99.99 | 1325-1600 | S.S. | 68.2 | 0.33 | 102 | (1958) |
| P | 1250-1725 | S.D. | 66.5 | 0.22 | 103 | (1962) |

POTASSIUM        Tracer $^{42}$K

| | | | | | | |
|---|---|---|---|---|---|---|
| P 99.95 | 0-60 | S.S. | 9.75 | 0.31 | 104 | (1967) |

γ-PLUTONIUM        Tracer $^{238}$Pu

| | | | | | | |
|---|---|---|---|---|---|---|
| P | 190-310 | S.S. | 16.7 | $2.1 \times 10^{-5}$ | 105 | (1966) |

δ-PLUTONIUM        Tracer $^{238}$Pu

| | | | | | | |
|---|---|---|---|---|---|---|
| P | 350-440 | S.S. | 23.8 | $4.5 \times 10^{-3}$ | 106 | (1964) |

| Material | Temperature range (°C) | Form of analysis | Activation energy Q (kcal/g-atom) | Frequency factor, $D_0$ (cm²/sec) | Reference No. | Year |
|---|---|---|---|---|---|---|
| ε-PLUTONIUM | Tracer $^{240}$Pu | | | | | |
| P | 500-612 | R.A. | 18.5 | $2 \times 10^{-2}$ | 107 | (1968) |
| SELENIUM | Tracer $^{75}$Se | | | | | |
| P | 35-140 | | 11.7 | $1.4 \times 10^{-4}$ | 108 | (1957) |
| SILICON | Tracer $^{31}$Si | | | | | |
| S 99.99999 | 1225-1400 | S.S. | 110.0 | 1800.0 | 109 | (1966) |
| S 99.999999 | 1100-1300 | S.S. | 118.5 | 9000.0 | 110 | (1966) |
| SILVER | Tracer $^{110}$Ag | | | | | |
| P | 500-900 | S.S. | 45.95 | 0.895 | 111 | (1941) |
| S,P | 500-950 | | 46.0 | 0.9 | 112 | (1949) |
| P,S 99.99 | 500-875 | S.S.,S.D.*** | 45.95*** | 0.895 | 113 | (1951) |
| P 99.99 | 670-940 | S.S.*** | 45.50*** | 0.724 | 114 | (1952) |
| P,S | 640-905 | S.S. | 40.8 | 0.11 | 115 | (1952) |
| P | 725-925 | | 47.4 | 1.8 | 116 | (1953) |
| P | 700-860 | A.R.G. | 45.0 | 0.905 | 117 | (1954) |
| P | 750-925 | | 45.0 | 0.53 | 118 | (1955) |
| P | 700-900 | R.A. | 45.4 | 0.65 | 119 | (1955) |
| S | 550-900 | S.S. | 44.05 | 0.40 | 120 | (1955) |
| P | 707-880 | A.R.G. | 44.9 | 0.834 | 121 | (1955) |
| S 99.99 | 630-940 | S.S.*** | 44.09*** | 0.395 | 122 | (1956) |
| P | 660-740 | | 41.8 | 2.78 | 123 | (1956) |
| S | 715-940 | S.S.*** | 43.7*** | 0.27 | 124 | (1957) |
| P | 500-900 | | 44.6 | 0.86 | 125 | (1957) |
| P | 690-900 | R.A. | 45.2 | 0.62 | 126 | (1957) |
| P | 750-950 | S.S. | 41.53 | 0.094 | 127 | (1958) |
| P | 650-900 | | 44.8 | 1.08 | 128 | (1958) |
| P 99.99 | 456-792 | R.A. | 45.8 | 0.69 | 129 | (1960) |
| P 99.99 | 250-380 | | 43.5 | 0.34 | 130 | (1963) |
| P | 718-914 | | 48.8 | 1.27 | 131 | (1964) |
| S | 750-900 | S.S.*** | 44.4*** | 1.06 | 132 | (1966) |
| S 99.999 | 668-958 | R.A. | Nonlinear | | 133 | (1968) |
| P | 720-950 | R.A. | 43.39 | 0.278 | 134 | (1968) |

| Material | Temperature range (°C) | Form of analysis | Activation energy Q (kcal/g-atom) | Frequency factor, $D_0$ (cm²/sec) | Reference No. | Year |
|---|---|---|---|---|---|---|
| SODIUM | Tracer $^{22}$Na | | | | | |
| | | | | | | |
| P | 10-95 | S.S. | 10.45 | 0.242 | 135 | (1952) |
| P 99.99 | 0-98 | S.S. | 10.09 | 0.145 | 136 | (1966) |
| | | | | | | |
| SULFUR | Tracer $^{35}$S | | | | | |
| | | | | | | |
| S⊥c | 40-100 | | 3.08 | $8.3 \times 10^{-12}$ | 137 | (1951) |
| S‖c | 40-110 | | 78.0 | $1.7 \times 10^{36}$ | 137 | (1951) |
| | | | | | | |
| TELLURIUM | Tracer $^{127}$Te | | | | | |
| | | | | | | |
| S 99.9999⊥c | 300-400 | S.S. | 46.7 | $3.91 \times 10^4$ | 138 | (1967) |
| S 99.9999‖c | 300-400 | S.S. | 35.5 | $1.30 \times 10^2$ | 138 | (1967) |
| | | | | | | |
| TANTALUM | Tracer $^{182}$Ta | | | | | |
| | | | | | | |
| P | 1830-2530 | | 110(89.5) | 2(0.03) | 139 | (1953) |
| P | 1200-1350 | R.A. | 110 | 1300 | 140 | (1955) |
| P,S 99.996 | 1250-2200 | S.S. | 98.7 | 1.24 | 141 | (1965) |
| | | | | | | |
| α-THALLIUM | Tracer $^{204}$Tl | | | | | |
| | | | | | | |
| S⊥c 99.9 | 135-230 | S.S. | 22.6 | 0.4 | 142 | (1955) |
| S‖c 99.9 | 135-230 | S.S. | 22.9 | 0.4 | 142 | (1955) |
| | | | | | | |
| β-THALLIUM | Tracer $^{204}$Tl | | | | | |
| | | | | | | |
| S 99.9 | 230-280 | S.S. | 20.7 | 0.7 | 142 | (1955) |
| | | | | | | |
| α-THORIUM | Tracer $^{228}$Th | | | | | |
| | | | | | | |
| P | | | 71.6 | 395 | 143 | (1967) |

| Material | Temperature range (°C) | Form of analysis | Activation energy Q (kcal/g-atom) | Frequency factor, $D_0$ (cm²/sec) | Reference No. | Year |
|---|---|---|---|---|---|---|
| β-TIN | Tracer $^{113}$Sn, $^{123}$Sn | | | | | |
| S⊥c | 180-225 | S.S. | 10.5 | $8.4 \times 10^{-4}$ | 144 | (1949) |
| S∥c | 180-225 | S.S. | 5.9 | $3.7 \times 10^{-8}$ | 144 | (1949) |
| S⊥c | 180-223 | S.S. | 6.7 | $9.2 \times 10^{-8}$ | 145 | (1950) |
| S∥c | 180-223 | S.S. | 9.4 | $3.6 \times 10^{-6}$ | 145 | (1950) |
| P 99.9 | −2-100 | A.R.G. | 9.4 | − | 15 | (1959) |
| S⊥c | 178-222 | S.S.*** | 23.3*** | 1.4 | 146 | (1960) |
| S∥c | 178-222 | S.S.*** | 25.6*** | 8.2 | 146 | (1960) |
| S,P | 140-217 | R.A. | 10.8 | $9.9 \times 10^{-4}$ | 147 | (1960) |
| P 99.99 | 130-255 | | 22.4 | 0.78 | 148 | (1961) |
| S⊥c 99.999 | 160-226 | S.S.*** | 25.1*** | 10.7 | 149 | (1964) |
| S∥c 99.999 | 160-226 | S.S.*** | 25.6*** | 7.7 | 149 | (1964) |
| P | 416-490 | | 7.7 | $1.73 \times 10^{-6}$ | 150 | (1967) |
| α-TITANIUM | Tracer $^{44}$Ti | | | | | |
| P | 690-850 | R.A. | 29.3 | $6.4 \times 10^{-8}$ | 151 | (1963) |
| β-TITANIUM | Tracer $^{44}$Ti | | | | | |
| P 99.95 | 900-1540 | S.S.*** | 31.2, 60.0*** | $3.58 \times 10^{-4}$, 1.09 | 152 | (1964) |
| P 99.9 | 900-1580 | R.A. | 36.5 | $1.9 \times 10^{-3}$ | 153 | (1968) |
| TUNGSTEN | Tracer $^{185}$W | | | | | |
| P | 1290-1450 | S.D. | 135.8 | $6.3 \times 10^{7}$ | 154 | (1950) |
| P | 2000-2700 | | 120.5 | 0.54 | 155 | (1961) |
| S 99.2 | 2670-3225 | S.S. | 153.1 | 42.8 | 156 | (1965) |
| P 99.9 | 1740-2100 | S.S. | 93.1 | $1.8 \times 10^{-3}$ | 157 | (1966) |
| α-URANIUM | Tracer $^{234}$U | | | | | |
| P | 580-650 | | 40.0 | $2 \times 10^{-3}$ | 158 | (1961) |
| β-URANIUM | Tracer $^{234}$U | | | | | |
| P | 700-760 | R.A. | 42.0 | 0.0135 | 159 | (1959) |

| Material | Temperature range (°C) | Form of analysis | Activation energy Q (kcal/g-atom) | Frequency factor, $D_0$ (cm²/sec) | Reference No. | Year |
|---|---|---|---|---|---|---|
| γ -URANIUM | Tracer $^{233}$U, $^{234}$U, $^{235}$U | | | | | |
| | | | | | | |
| P | 800-1050 | A.R.G. | 21.0 | $1.4 \times 10^{-4}$ | 160 | (1958) |
| P | 800-1050 | S.S. | 26.6 | $1.17 \times 10^{-3}$ | 160 | (1958) |
| P | 800-1040 | S.S. | 27.5 | $1.8 \times 10^{-3}$ | 161 | (1958) |
| P 99.99 | 800-1070 | S.S. | 28.5 | $2.33 \times 10^{-3}$ | 162 | (1960) |
| | | | | | | |
| VANADIUM | Tracer $^{48}$V | | | | | |
| | | | | | | |
| S 99.7 | 1000-1400 | S.S. | 61.0 | $1.1 \times 10^{-2}$ | 163 | (1965) |
| S 99.7 | 1600-1890 | S.S. | 91.5 | 58.0 | 163 | (1965) |
| S,P 99.99 | 880-1360 | S.S.*** | 73.65*** | 0.36 | 164 | (1965) |
| S,P 99.99 | 1360-1830 | S.S.*** | 94.14*** | 214.0 | 164 | (1965) |
| S 99.92 | 700-1050 | R.A. | 64.6 | 0.107 | 165 | (1968) |
| S 99.92 | 1050-1400 | R.A. | 76.8 | 10.45 | 165 | (1968) |
| | | | | | | |
| ZINC | Tracer $^{65}$Zn | | | | | |
| | | | | | | |
| S⊥c | 340-410 | S.S. | 19.6 | 0.02 | 166 | (1941) |
| S⊥c | 340-410 | S.S. | 31.0 | 93.0 | 167 | (1942) |
| S‖c | 340-410 | S.S. | 20.4 | 0.05 | 167 | (1942) |
| S⊥c 99.999 | 300-400 | S.S. | 25.4 | 1.3 | 168 | (1952) |
| S‖c 99.999 | 300-400 | S.S. | 21.7 | 0.1 | 168 | (1952) |
| S⊥c 99.999 | 240-400 | S.S. | 24.3 | 0.58 | 169 | (1953) |
| S‖c 99.999 | 240-400 | S.S. | 21.8 | 0.13 | 169 | (1953) |
| S⊥c | 260-400 | S.S. | 25.9 | 1.6 | 170 | (1954) |
| S‖c | 260-400 | S.S. | 19.6 | 0.02 | 170 | (1954) |
| P | 260-400 | S.S. | 23.8 | 0.41 | 170 | (1954) |
| S⊥c | 200-415 | S.S.,S.D.*** | 24.9*** | 0.39 | 171 | (1956) |
| S‖c | 200-415 | S.S.,S.D.*** | 22.0*** | 0.08 | 171 | (1956) |
| P | 200-415 | S.S.,S.D. | 22.7 | 0.19 | 171 | (1956) |
| S⊥c | 270-370 | | 26.0 | 2.78 | 172 | (1957) |
| S‖c | 270-370 | | 19.0 | 0.013 | 172 | (1957) |
| P | 325-405 | S.S. | 20.5 | 0.031 | 173 | (1959) |
| S⊥c 99.999 | 240-418 | S.S.*** | 23.0*** | 0.18 | 174 | (1967) |
| S‖c 99.999 | 240-418 | S.S.*** | 21.9*** | 0.13 | 174 | (1967) |

| Material | Temperature range (°C) | Form of analysis | Activation energy Q (kcal/g-atom) | Frequency factor, $D_0$ (cm²/sec) | Reference No. | Year |
|---|---|---|---|---|---|---|
| $\alpha$-ZIRCONIUM | Tracer $^{95}$Zr | | | | | |
| P 99.96 | 700-800 | R.A. | 22.0 | $5 \times 10^{-8}$ | 175 | (1958) |
| P 99.96 | 650-825 | R.A. | 52.0 | $5.9 \times 10^{-2}$ | 176 | (1959) |
| P 99.95 | 750-850 | S.S. | 45.5 | $5.6 \times 10^{-4}$ | 177 | (1963) |
| $\beta$-ZIRCONIUM | Tracer $^{95}$Zr | | | | | |
| P 99.9 | 1000-1250 | R.A. | 26.0 | $4.0 \times 10^{-5}$ | 178 | (1959) |
| P 99.6 | 1115-1500 | R.A. | 38.0 | $2.4 \times 10^{-3}$ | 176 | (1959) |
| P 99.6 | 900-1200 | R.A. | 27.0 | $1 \times 10^{-4}$ | 179 | (1960) |
| P | 900-1240 | S.S. | 24.0 | $4.2 \times 10^{-5}$ | 180 | (1960) |
| P 99.89 | 1100-1500 | S.S. | 30.1 | $2.4 \times 10^{-4}$ | 181 | (1961) |
| P 99.94 | 900-1750 | S.S.*** | 19.6 +*** $30.9(T-1136)$ | $3 \times 10^{-6} \times$ $(T/1136)^{15.6}$ | 182 | (1963) |

# Part II

## Impurity Diffusion in Pure Metals

| Solute | Material | Temperature range (°C) | Form of analysis | Activation energy, Q (kcal/mole) | Frequency factor, $D_0$ (cm²/sec) | Reference No. | Year |
|--------|----------|------------------------|------------------|----------------------------------|-----------------------------------|---------------|------|
| ALUMINUM | | | | | | | |
| $^{110}$Ag | P | 400-630 | R.A. | 29.0 | 0.21 | 183 | (1967) |
| $^{110}$Ag | P | 450-632 | R.A. | 27.9 | 0.08 | 184 | (1968) |
| $^{115}$Cd | P 99.995 | 400-630 | R.A. | 21.7 | $7.94 \times 10^{-8}$ | 183 | (1967) |
| $^{64}$Cu | P | 350-630 | R.A. | 30.20 | 0.15 | 185 | (1965) |
| $^{51}$Cr | P 99.99 | 360-630 | R.A. | 19.9 | $1.1 \times 10^{-6}$ | 186 | (1962) |
| $^{51}$Cr | P 99.95 | 250-605 | R.A. | 15.4 | $3.01 \times 10^{-7}$ | 187 | (1964) |
| $^{59}$Fe | P 99.99 | 360-630 | R.A. | 13.9 | $4.1 \times 10^{-9}$ | 186 | (1962) |
| $^{114}$In | P 99.995 | 400-630 | R.A. | 22.2 | $1.42 \times 10^{-8}$ | 183 | (1967) |
| $^{54}$Mn | P 99.99 | 450-650 | S.S. | 28.8 | 0.22 | 1 | (1962) |
| $^{99}$Mo | P 99.995 | 400-630 | R.A. | 13.1 | $1.04 \times 10^{-9}$ | 188 | (1967) |
| $^{63}$Ni | P 99.99 | 360-630 | R.A. | 15.7 | $2.9 \times 10^{-8}$ | 186 | (1962) |
| $^{103}$Pd | P 99.995 | 400-630 | R.A. | 20.2 | $1.92 \times 10^{-7}$ | 183 | (1967) |
| $^{124}$Sb | P | 448-620 | R.A. | 29.1 | 0.09 | 184 | (1968) |
| $^{113}$Sn | P 99.995 | 400-630 | R.A. | 20.2 | $3.05 \times 10^{-7}$ | 183 | (1967) |
| $^{48}$V | P 99.995 | 400-630 | R.A. | 19.6 | $6.05 \times 10^{-8}$ | 189 | (1968) |
| $^{65}$Zn | P 99.99 | 405-654 | S.S. | 30.9 | 1.1 | 173 | (1959) |
| $^{65}$Zn | S 99.995 | 327-375 | R.A. | 30.8 | 1.4 | 190 | (1960) |
| BERYLLIUM | | | | | | | |
| $^{110}$Ag | ‖c 99.75 | 656-897 | R.A. *** | 39.1 *** | 0.41 | 191 | (1964) |
| $^{110}$Ag | ⊥c 99.75 | 656-897 | R.A. *** | 45.7 *** | 1.98 | 191 | (1964) |
| $^{110}$Ag | P 99.75 | 656-897 | R.A. | 46.1 | 6.2 | 191 | (1964) |

43

| Solute | Material | Temperature range (°C) | Form of analysis | Activation energy, Q (kcal/mole) | Frequency factor, $D_0$ (cm²/sec) | Reference No. | Year |
|---|---|---|---|---|---|---|---|
| **BERYLLIUM (continued)** | | | | | | | |
| $^{110}$Ag | S⊥c 99.75 | 650-900 | R.A. *** | 43.2 *** | 1.76 | 192 | (1965) |
| $^{110}$Ag | S∥c 99.75 | 650-900 | R.A. *** | 39.3 *** | 0.43 | 192 | (1965) |
| $^{59}$Fe | S 99.75 | 700-1076 | R.A. | 51.6 | 0.67 | 192 | (1965) |
| $^{55}$Fe | c 99.75 | 656-897 | R.A. | 51.8 | 0.53 | 191 | (1964) |
| **CADMIUM** | | | | | | | |
| $^{110}$Ag | S 99.99 | 180-300 | — | 25.4 | 2.21 | 10 | (1967) |
| $^{65}$Zn | S 99.99 | 180-300 | — | 19.0 | 0.0016 | 10 | (1967) |
| **CARBON** | | | | | | | |
| $^{110}$Ag | ⊥c | 750-1050 | R.A. | 64.3 | $9.28 \times 10^3$ | 193 | (1965) |
| $^{63}$Ni | ⊥c | 540-920 | R.A. | 47.2 | 102 | 193 | (1965) |
| $^{63}$Ni | ∥c | 750-1060 | R.A. | 53.3 | 2.2 | 193 | (1965) |
| $^{228}$Th | ⊥c | 1400-2200 | R.A. | 145.4 | $1.33 \times 10^5$ | 193 | (1965) |
| $^{228}$Th | ∥c | 1800-2200 | R.A. | 114.7 | 2.48 | 193 | (1965) |
| $^{228}$Th | | 1600-2000 | S.S. | 114 | 15.5 | 194 | (1967) |
| $^{232}$U | ⊥c | 1400-2200 | R.A. | 115 | $6.76 \times 10^3$ | 193 | (1965) |
| $^{232}$U | ∥c | 1400-1820 | R.A. | 129.5 | 385 | 193 | (1965) |
| **CHROMIUM** | | | | | | | |
| $^{55}$Fe | P 99.8 | 980-1420 | R.A. | 79.3, 40.4 | 0.47, $1.1 \times 10^{-6}$ | 195 | (1964) |
| **COBALT** | | | | | | | |
| $^{14}$C | P 99.82 | 600-1400 | — | 34.0 | 0.21 | 196 | (1963) |
| $^{55}$Fe | P 99.9 | 1104-1303 | S.S. | 62.7 | 0.21 | 23 | (1955) |
| $^{55}$Fe | P 99.97 | 930-1240 | S.D. | 55.5 | 0.2 | 197 | (1965) |
| $^{63}$Ni | P 99.5 | 1152-1400 | S.D. | 72.1 | 1.25 | 85 | (1959) |
| $^{63}$Ni | P | 772-1048 | R.A. *** | 64.3 *** | 0.34 | 24 | (1962) |
| $^{63}$Ni | P | 1192-1297 | R.A. *** | 60.2 *** | 0.10 | 24 | (1962) |
| $^{63}$Ni | P 99.5 | 1227-1416 | R.A. | 71.0 | 3.35 | 25 | (1965) |
| $^{35}$S | P 99.99 | 1150-1250 | R.A. | 5.4 | 1.3 | 198 | (1964) |

| Solute | Material | Temperature range (°C) | Form of analysis | Activation energy, Q (kcal/mole) | Frequency factor, $D_0$ (cm²/sec) | Reference No. | Year |
|--------|----------|----------------------|-----------------|------------------------------|------------------------------|------|------|
| **COPPER** | | | | | | | |
| $^{110}$Ag | | − | − | 46.5 | 0.63 | 199 | (1960) |
| $^{76}$As | | − | − | 42.0 | 0.12 | 199 | (1960) |
| $^{198}$Au | P 99.95 | 750–1000 | S.S. | 44.9 | 0.1 | 200 | (1954) |
| $^{198}$Au | | − | − | 49.7 | 0.69 | 199 | (1960) |
| $^{198}$Au | P 99.999 | 706–1010 | S.S. *** | 45.7 *** | 0.15 | 46 | (1965) |
| $^{115}$Cd | S 99.98 | 725–950 | S.S. | 45.7 | 0.935 | 201 | (1958) |
| $^{60}$Co | S 99.99 | 700–950 | S.S. | 55.2 | 5.7 | 202 | (1958) |
| $^{60}$Co | S 99.998 | 701–1077 | S.S. | 54.1 | 1.93 | 203 | (1958) |
| $^{55}$Fe | S 99.99 | 1130–1320 | S.S. | 93.02 | $1.6 \times 10^6$ | 204 | (1958) |
| $^{59}$Fe | S 99.998 | 720–1075 | S.S. | 51.8 | 1.4 | 203 | (1958) |
| $^{59}$Fe | S 99.99 | 720–1060 | S.S. | 50.95 | 1.01 | 205 | (1961) |
| $^{72}$Ga | | − | − | 45.90 | 0.55 | 199 | (1960) |
| $^{203}$Hg | | − | − | 44.0 | 0.35 | 199 | (1960) |
| $^{54}$Mn | S 99.99 | 754–950 | S.S. | 91.4 | $10^7$ | 206 | (1959) |
| $^{63}$Ni | P | 650–915 | A.R.G. | 64.8 | | 207 | (1955) |
| $^{63}$Ni | S 99.998 | 743–1076 | S.S. *** | 56.5 *** | 2.7 | 203 | (1958) |
| $^{63}$Ni | S 99.99 | 695–1061 | S.S. *** | 56.8 *** | 3.8 | 208 | (1959) |
| $^{63}$Ni | P 99.9999 | 250–520 | S.D. | 46.8 | $4.2 \times 10^{-2}$ | 209 | (1966) |
| $^{63}$Ni | S | 899–1067 | S.S. *** | 55.3 *** | 1.7 | 210 | (1968) |
| $^{63}$Ni | P | 300–450 | S.D. | 45.3 | $6.4 \times 10^{-2}$ | 211 | (1968) |
| $^{66}$Ni | S 99.999 | 855–1055 | S.S. *** | 55.6 *** | 1.93 | 212 | (1968) |
| $^{102}$Pd | S 99.999 | 807–1056 | S.S. | 54.37 | 1.71 | 213 | (1963) |
| $^{195}$Pt | P | 843–997 | S.S. | 37.5 | $4.8 \times 10^{-4}$ | 214 | (1963) |
| $^{124}$Sb | S 99.999 | 600–1000 | S.S. | 42.0 | 0.34 | 215 | (1960) |
| $^{113}$Sn | | 680–910 | | 45.0 | 0.11 | 216 | (1962) |
| $^{204}$Tl | S 99.999 | 785–996 | S.S. | 43.3 | 0.71 | 217 | (1963) |
| $^{65}$Zn | S | 600–1050 | S.S. | 45.6 | 0.34 | 218 | (1957) |
| $^{65}$Zn | | 800–850 | | 45 | − | 219 | (1966) |
| | | | | | | | |
| **GERMANIUM** | | | | | | | |
| $^{115}$Cd | S | 750–950 | R.A. | 102.0 | $1.75 \times 10^9$ | 220 | (1959) |
| $^{59}$Fe | S | 775–930 | R.A. | 24.8 | 0.13 | 221 | (1957) |
| $^{114}$In | S | 600–920 | − | 39.9 | $2.9 \times 10^{-4}$ | 222 | (1961) |
| $^{124}$Sb | S | 800–900 | − | 52 | 1.3 | 223 | (1957) |
| $^{124}$Sb | S | 720–900 | − | 50.2 | 0.22 | 224 | (1964) |
| $^{125}$Te | S | 770–900 | S.S. | 65, 56 | 2, 6 | 225 | (1962) |
| $^{204}$Tl | S | 800–930 | S.S. | 78.4 | 1700 | 226 | (1962) |

| Solute | Material | Temperature range (°C) | Form of analysis | Activation energy, Q (kcal/mole) | Frequency factor, $D_0$ (cm²/sec) | Reference No. | Year |
|--------|----------|------------------------|------------------|----------------------------------|-----------------------------------|---------------|------|
| GOLD | | | | | | | |
| $^{110}$Ag | S 99.99 | 699-1007 | S.S. | 40.2 | 0.072 | 227 | (1963) |
| $^{110}$Ag | P 99.99 | 773-1039 | – | 40.4 | 0.08 | 228 | (1965) |
| $^{60}$Co | P 99.93 | 702-948 | R.A. | 41.6 | 0.068 | 44 | (1963) |
| $^{59}$Fe | P 99.93 | 701-948 | R.A. | 41.6 | 0.082 | 44 | (1963) |
| $^{203}$Hg | S 99.994 | 600-1027 | – | 37.38 | 0.116 | 229 | (1965) |
| $^{63}$Ni | P 99.96 | 880-940 | S.S. | 46.0 | 0.30 | 83 | (1957) |
| $^{63}$Ni | P 99.93 | 702-988 | R.A. | 42.0 | 0.034 | 44 | (1963) |
| $^{195}$Pt | P,S 99.98 | 800-1060 | S.S. | 60.9 | 7.6 | 230 | (1960) |
| INDIUM | | | | | | | |
| $^{110}$Ag | S⊥c 99.99 | 25-140 | S.S. | 12.8 | 0.52 | 231 | (1966) |
| $^{110}$Ag | S∥c 99.99 | 25-140 | S.S. | 11.5 | 0.11 | 231 | (1966) |
| $^{198}$Au | S 99.99 | 25-140 | S.S. | 6.7 | $9 \times 10^{-3}$ | 231 | (1966) |
| $^{204}$Tl | S 99.99 | 49-157 | S.S. | 15.5 | 0.049 | 48 | (1952) |
| α-IRON | | | | | | | |
| $^{195}$Au | P 99.999 | 800-900 | R.A. | 62.4 | 31 | 232 | (1963) |
| $^{14}$C | P | 350-850 | R.A. | 19.2 | $6.2 \times 10^{-3}$ | 233 | (1955) |
| $^{14}$C | P | 500-800 | R.A. | 24.6 | 0.2 | 234 | (1954) |
| $^{14}$C | P 99.98 | 616-844 | R.A. | 29.3 | 2.2 | 235 | (1964) |
| $^{60}$Co | P | 700-790 | S.D. | 54.0 | 0.2 | 236 | (1954) |
| $^{60}$Co | P 99.5 | 700-850 | S.D. | 54.0 | 0.4 | 237 | (1954) |
| $^{60}$Co | P | 800-905 | S.D. | 64.6 | 64.4 | 238 | (1961) |
| $^{60}$Co | P 99.999 | 690-905 | R.A. | 68.3 | 118 | 234 | (1963) |
| $^{60}$Co | P 99.98 | 825-890 | S.S. *** | 62.3 *** | 9.5 | 239 | (1964) |
| $^{60}$Co | P 99.995 | 638-768 | R.A. *** | 62.2 *** | 7.19' | 62 | (1966) |
| $^{60}$Co | P 99.995 | 808-884 | R.A. *** | 61.4 *** | 6.38 | 62 | (1966) |
| $^{51}$Cr | P 99.5 | 750-850 | S.D. | 82.0 | $3 \times 10^4$ | 237 | (1954) |
| $^{51}$Cr | 99.95 | 775-875 | R.A. *** | 57.5 *** | 2.53 | 240 | (1967) |
| $^{64}$Cu | P 99.0 | 800-1050 | R.A. | 57.0 | 0.57 | 241 | (1966) |
| $^{64}$Cu | P 99.0 | 650-750 | R.A. | 58.38 | 0.47 | 241 | (1966) |
| $^{42}$K | P 99.92 | 500-800 | R.A. | 42.3 | 0.036 | 242 | (1967) |
| $^{42}$K | P 99.92 | 500-800 | R.A. | 30.6 | $4.6 \times 10^{-4}$ | 242 | (1967) |
| $^{42}$K | P 99.92 | 600-800 | R.A. | 24.2 | $1.17 \times 10^{-7}$ | 242 | (1967) |
| $^{99}$Mo | P | 750-875 | R.A. | 73.0 | 7800 | 243 | (1966) |
| $^{63}$Ni | P 99.97 | 600-680 | R.A.,S.D. | 58.7 | 1.4 | 244 | (1961) |

| Solute | Material | Temperature range (°C) | Form of analysis | Activation energy, Q (kcal/mole) | Frequency factor, $D_0$ (cm²/sec) | Reference No. | Year |
|--------|----------|------------------------|------------------|----------------------------------|------------------------------------|---------------|------|
| α-IRON | (continued) | | | | | | |
| $^{63}$Ni | P 99.97 | 680-800 | R.A.,S.D. | 56.0 | 1.3 | 244 | (1961) |
| $^{63}$Ni | P 99.999 | 700-900 | R.A. | 61.9 | 9.9 | 232 | (1963) |
| $^{32}$P | P 99.99 | 700-850 | — | 40.0 | $7.1 \times 10^{-3}$ | 245 | (1963) |
| $^{32}$P | P | 860-900 | R.A. | 55.0 | 2.9 | 246 | (1963) |
| $^{124}$Sb | P | 800-900 | R.A. | 66.6 | 1100 | 247 | (1968) |
| $^{185}$W | P 99.5 | 700-785 | S.D. | 70.0 | 380 | 237 | (1954) |
| $^{185}$W | P 99.99 | 700-900 | R.A. | 63.5 | 69 | 248 | (1967) |
| γ -IRON | | | | | | | |
| $^{7}$Be | P 99.9 | 1100-1350 | R.A. | 57.6 | 0.1 | 249 | (1968) |
| $^{14}$C | P | 900-1050 | R.A. | 32.4 | 0.1 | 234 | (1955) |
| $^{14}$C | P 99.34 | 800-1400 | — | 34.0 | 0.15 | 196 | (1963) |
| $^{60}$Co | P 99.5 | 1050-1250 | R.A. | 104.0 | $1.2 \times 10^5$ | 237 | (1954) |
| $^{60}$Co | P | 1100-1200 | — | 87.0 | 300 | 250 | (1955) |
| $^{60}$Co | P 99.98 | 1138-1340 | S.S. *** | 72.9 *** | 1.25 | 251 | (1961) |
| $^{51}$Cr | P 99.5 | 1050-1250 | R.A. | 97.0 | $1.8 \times 10^4$ | 237 | (1954) |
| $^{64}$Cu | P 99.2 | 800-1200 | R.A. | 61.0 | 3 | 252 | (1955) |
| $^{181}$Hf | P 99.94 | 1075-1340 | R.A. | 113.0 | $9 \times 10^4$ | 68 | (1965) |
| $^{95}$Nb | P 99.94 | 1075-1340 | R.A. | 82.3 | 530 | 68 | (1965) |
| $^{63}$Ni | P 99.91 | 1152-1400 | S.D. | 77.6 | 6.92 | 85 | (1959) |
| $^{63}$Ni | P 99.97 | 930-1050 | S.D.,R.A. | 67.0 | 0.77 | 244 | (1961) |
| $^{32}$P | P | 1280-1350 | R.A. | 69.8 | 28.3 | 246 | (1963) |
| $^{32}$P | P 99.99 | 950-1200 | R.A. | 43.7 | $1.0 \times 10^{-2}$ | 245 | (1963) |
| $^{35}$S | P 99.97 | 1200-1350 | S.D. | 48.4 | 1.35 | 254 | (1962) |
| $^{185}$W | P 99.5 | 1050-1250 | R.A. | 90.0 | 1000 | 239 | (1954) |
| δ -IRON | | | | | | | |
| $^{60}$Co | P 99.98 | 1396-1502 | S.S.,R.A. | 61.2 | 5.5 | 70 | (1963) |
| $^{60}$Co | P 99.995 | 1428-1521 | R.A. | 61.4 | 6.38 | 62 | (1966) |
| $^{32}$P | P 99.99 | 1370-1460 | R.A. | 55.0 | 2.9 | 246 | (1963) |
| LEAD | | | | | | | |
| $^{110}$Ag | S,P | 220-320 | S.S. | 8.02 | $7.9 \times 10^{-3}$ | 254 | (1966) |
| $^{195}$Au | S,P 99.999 | 94-325 | S.S. | 9.35 | $4.1 \times 10^{-3}$ | 255 | (1960) |
| $^{198}$Au | S 99.999 | 190-320 | S.S. | 10.0 | $8.7 \times 10^{-3}$ | 256 | (1966) |

| Solute | Material | Temper-ature range (°C) | Form of analy-sis | Activation energy, Q (kcal/mole) | Frequency factor, $D_0$ (cm²/sec) | Reference No. | Year |
|---|---|---|---|---|---|---|---|
| **LEAD** (continued) | | | | | | | |
| $^{115}$Cd | | 150-330 | | 21.2 | 0.405 | 257 | (1967) |
| $^{64}$Cu | S,P | 150-320 | S.S. | 14.44 | $4.6 \times 10^{-2}$ | 254 | (1966) |
| $^{205}$Tl | P 99.999 | 207-322 | S.S. | 24.33 | 0.511 | 258 | (1961) |
| **LITHIUM** | | | | | | | |
| $^{22}$Na | P | 52-176 | S.S. | 12.61 | 0.41 | 259 | (1967) |
| $^{110}$Ag | P | 67-160 | S.S. | 12.83 | 0.37 | 260 | (1968) |
| **MOLYBDENUM** | | | | | | | |
| $^{14}$C | P 99.98 | 1200-1600 | R.A. | 41.0 | $2.04 \times 10^{-2}$ | 261 | (1966) |
| $^{60}$Co | P 99.98 | 1900-2300 | A.R.G. | 100 | 3 | 262 | (1962) |
| $^{60}$Co | P 99.98 | 1850-2350 | S.S. *** | 106.7 | 18 | 263 | (1965) |
| $^{42}$K | S | 800-1100 | S.S. | 26.5 | $9.34 \times 10^{-9}$ | 264 | (1966) |
| $^{42}$K | P | 800-1100 | S.S. | 14.6 | $2.86 \times 10^{-10}$ | 264 | (1966) |
| $^{95}$Nb | P 99.98 | 1850-2350 | S.S. | 108.1 | 14 | 263 | (1965) |
| $^{186}$Re | P | 1700-2100 | A.R.G. | 94.7 | $9.7 \times 10^{-2}$ | 265 | (1965) |
| $^{35}$S | S 99.97 | 2220-2470 | S.S. | 101.0 | 320 | 266 | (1968) |
| $^{235}$U | P 99.98 | 1500-2000 | R.A. | 76.4 | $7.6 \times 10^{-3}$ | 267 | (1965) |
| $^{185}$W | P | 1750-2150 | R.A. | 78 | $4 \times 10^{-4}$ | 77 | (1959) |
| $^{186}$W | P | 1700-2100 | A.R.G. *** | 112.9*** | 3.18 | 265 | (1965) |
| $^{185}$W | P 99.98 | 1700-2260 | S.S.*** | 110*** | 1.7 | 268 | (1967) |
| $^{185}$W | S | 2000-2430 | S.S. | 151 | 25.1 | 269 | (1966) |
| **NICKEL** | | | | | | | |
| $^{198}$Au | P | 900-1060 | A.R.G. | 65.0 | 2.0 | 40 | (1955) |
| $^{198}$Au | S,P 99.999 | 700-1075 | S.S. | 55.0 | 0.02 | 270 | (1968) |
| $^{7}$Be | P 99.9 | 1020-1400 | R.A. | 46.2 | 0.019 | 249 | (1968) |
| $^{14}$C | P | 500-800 | R.A. | 33 | 0.08 | 271 | (1957) |
| $^{14}$C | P 99.86 | 600-1400 | — | 34.0 | 0.12 | 196 | (1953) |
| $^{60}$Co | P 99.9 | 900-1250 | S.D. | 68.3 | 1.46 | 272 | (1951) |
| $^{60}$Co | P 99.8 | 748-1192 | R.A. | 64.7 | 0.75 | 24 | (1962) |
| $^{60}$Co | P 99.97 | 1149-1390 | R.A. *** | 65.9*** | 1.39 | 25 | (1965) |
| $^{51}$Cr | P 99.997 | 350-600 | R.A. | 13.7 | $5.45 \times 10^{-9}$ | 273 | (1964) |
| $^{51}$Cr | P 99.997 | 600-900 | R.A. | 40.8 | 0.03 | 273 | (1964) |
| $^{51}$Cr | P 99.95 | 1100-1270 | S.S. | 65.1 | 1.1 | 32 | (1964) |
| $^{64}$Cu | P 99.95 | 850-1050 | R.A. | 61.0 | 0.724 | 185 | (1965) |

| Solute | Material | Temperature range (°C) | Form of analysis | Activation energy, Q (kcal/mole) | Frequency factor, $D_0$ (cm²/sec) | Reference No. | Year |
|---|---|---|---|---|---|---|---|
| **NICKEL (continued)** | | | | | | | |
| $^{64}$Cu | P 99.95 | 1050–1360 | S.S. | 61.7 | 0.57 | 32 | (1964) |
| $^{59}$Fe | P | 950–1150 | R.A. | 51.0 | $8.4 \times 10^{-3}$ | 274 | (1953) |
| $^{59}$Fe | P | 400–800 | – | 31.8 | $7.3 \times 10^{-4}$ | 275 | (1955) |
| $^{59}$Fe | P 99.99 | 940–1100 | R.A. | 61.0 | 0.8 | 276 | (1960) |
| $^{63}$Ni | P 99.99 | 900–1200 | R.A. | 69.2 | 2.22 | 277 | (1966) |
| $^{63}$Ni | S 99.99 | 900–1200 | R.A. | 70.1 | 2.59 | 277 | (1966) |
| $^{124}$Sb | P 99.97 | 1020–1220 | – | 27.0 | $1.8 \times 10^{-5}$ | 278 | (1962) |
| $^{113}$Sn | P | 700–1000 | A.R.G. | 21.0 | $2.1 \times 10^{-7}$ | 279 | (1956) |
| $^{113}$Sn | P 99.8 | 700–1350 | A.R.G. | 58.0 | 0.83 | 15 | (1959) |
| $^{113}$Sn | P | 1270–1480 | R.A. | 65.5 | 30.0 | 280 | (1967) |
| $^{48}$V | P 99.99 | 800–1300 | R.A. | 66.50 | 0.87 | 189 | (1968) |
| $^{185}$W | P,S 99.9 | 1100–1275 | S.S. | 71.0 | 1.13 | 281 | (1958) |
| $^{185}$W | P 99.95 | 1100–1300 | S.S. | 71.5 | 2.0 | 185 | (1964) |
| **NIOBIUM** | | | | | | | |
| $^{14}$C | | 900–1100 | | 32.0 | $1.09 \times 10^{-5}$ | 282 | (1961) |
| $^{14}$C | P 99.14 | 1100–1400 | R.A. | 35.0 | $9.32 \times 10^{-2}$ | 261 | (1966) |
| $^{14}$C | P 99.5 | 930–1800 | R.A. *** | 37.9 *** | 0.033 | 283 | (1967) |
| $^{60}$Co | P 99.85 | 1500–2100 | A.R.G. | 70.5 | 0.74 | 262 | (1962) |
| $^{51}$Cr | S | 953–1435 | S.S. | 83.5 | 0.30 | 284 | (1968) |
| $^{51}$Cr | | 947–1493 | S.S. | 80.59 | 0.13 | 284 | (1968) |
| $^{55}$Fe | P 99.85 | 1400–2100 | A.R.G. | 77.7 | 1.5 | 262 | (1962) |
| $^{95}$Nb | | 1600–2000 | | 113 | 28.18 | 285 | (1967) |
| $^{35}$S | S 99.9 | 1100–1500 | R.A. | 73.1 | 2600 | 286 | (1968) |
| $^{113}$Sn | P 99.85 | 1850–2400 | S.S. | 78.9 | 0.14 | 287 | (1965) |
| $^{182}$Ta | P,S 99.997 | 878–2395 | S.S. | 99.3 | 1.0 | 98 | (1965) |
| $^{235}$U | P 99.55 | 1500–2000 | R.A. | 76.8 | $8.9 \times 10^{-3}$ | 267 | (1965) |
| $^{48}$V | S 99.99 | 1000–1400 | R.A. | 85.0 | 2.21 | 165 | (1968) |
| **PALLADIUM** | | | | | | | |
| $^{59}$Fe | P | 1240–1450 | R.A. | – | – | 280 | (1967) |
| **PLATINUM** | | | | | | | |
| $^{64}$Cu | P – | 1098–1375 | S.S. | 59.5 | 0.074 | 214 | (1963) |
| **POTASSIUM** | | | | | | | |
| $^{22}$Na | | 0–62 | S.S. | 7.45 | 0.058 | 288 | (1967) |

| Solute | Material | Temperature range (°C) | Form of analysis | Activation energy, Q (kcal/mole) | Frequency factor, $D_0$ (cm²/sec) | Reference No. | Year |
|---|---|---|---|---|---|---|---|

SELENIUM

| Solute | Material | Temperature range (°C) | Form of analysis | Activation energy, Q (kcal/mole) | Frequency factor, $D_0$ (cm²/sec) | Reference No. | Year |
|---|---|---|---|---|---|---|---|
| $^{59}$Fe | P — | 40-100 | R.A. | 8.88 | — | 289 | (1958) |
| $^{203}$Hg | P 99.996 | 25-100 | R.A. | 1.2 | — | 289 | (1958) |
| $^{95}$Zr | P 99.996 | 40-100 | R.A. | 6.3 | — | 289 | (1958) |

SILICON

| Solute | Material | Temperature range (°C) | Form of analysis | Activation energy, Q (kcal/mole) | Frequency factor, $D_0$ (cm²/sec) | Reference No. | Year |
|---|---|---|---|---|---|---|---|
| $^{198}$Au | S — | 700-1300 | R.A. | 25.8 | $1.1 \times 10^{-3}$ | 290 | (1956) |
| $^{198}$Au | S — | 700-1300 | S.S. *** | 8.9*** | $2.4 \times 10^{-4}$ | 291 | (1964) |
| $^{198}$Au | S — | 700-1300 | S.S. *** | 47.0*** | $2.75 \times 10^{-3}$ | 291 | (1964) |
| $^{14}$C | P — | 1070-1400 | R.A. | 67.2 | 0.33 | 292 | (1961) |
| $^{64}$Cu | P — | 800-1100 | R.A. | 23.0 | $4 \times 10^{-2}$ | 293 | (1958) |
| $^{59}$Fe | S — | 1000-1200 | R.A. | 20.0 | $6.2 \times 10^{-3}$ | 290 | (1956) |
| $^{63}$Ni |  | 450-800 |  | 97.5 | 1000 | 294 | (1967) |
| $^{32}$P | S — | 1100-1250 | — | 41.5 | — | 295 | (1962) |
| $^{124}$Sb | S — | 1190-1398 | R.A. | 91.7 | 12.9 | 296 | (1959) |

SILVER

| Solute | Material | Temperature range (°C) | Form of analysis | Activation energy, Q (kcal/mole) | Frequency factor, $D_0$ (cm²/sec) | Reference No. | Year |
|---|---|---|---|---|---|---|---|
| $^{198}$Au | S 99.99 | 650-950 | S.S. | 45.5 | 0.26 | 297 | (1956) |
| $^{198}$Au | P 99.99 | 656-905 | S.S. | 46.4 | 0.41 | 42 | (1957) |
| $^{198}$Au | S 99.99 | 718-942 | S.S. *** | 48.28 *** | 0.85 | 227 | (1963) |
| $^{115}$Cd | S 99.99 | 592-937 | S.S. | 41.687 | 0.441 | 298 | (1954) |
| $^{60}$Co | S 99.99 | 745-943 | S.S. | 59.9 | 104 | 299 | (1961) |
| $^{60}$Co | P 99.999 | 600-850 | R.A. | 29.9 | $3.0 \times 10^{-7}$ | 300 | (1963) |
| $^{64}$Cu | P 99.99 | 717-945 | S.S. | 46.1 | 1.23 | 301 | (1957) |
| $^{59}$Fe | P | 300-725 | — | 21.5 | $2.0 \times 10^{-5}$ | 252 | (1955) |
| $^{55}$Fe | S 99.99 | 720-930 | S.S. *** | 49.04*** | 2.42 | 205 | (1962) |
| $^{59}$Fe | P 99.999 | 600-850 | R.A. | 29.6 | $9.4 \times 10^{-7}$ | 300 | (1963) |
| $^{77}$Ge | P — | 640-870 | S.S. | 36.5 | 0.084 | 302 | (1958) |
| $^{203}$Hg | P 99.99 | 653-948 | S.S. | 38.1 | 0.079 | 301 | (1957) |
| $^{114}$In | S 99.99 | 592-937 | S.S. | 40.801 | 0.412 | 298 | (1954) |
| $^{114}$In | P | 770-940 |  | 41.79 | 0.55 | 303 | (1967) |
| $^{63}$Ni | S 99.99 | 749-950 | S.S. *** | 54.8 *** | 21.9 | 304 | (1961) |
| $^{63}$Ni | P 99.999 | 600-850 | R.A. | 28.7 | $8.5 \times 10^{-7}$ | 299 | (1963) |
| $^{210}$Pb | P — | 700-800 | S.S. | 39.2 | 0.39 | 305 | (1953) |
| $^{210}$Pb | P — | 700-865 | S.S. | 38.1 | 0.22 | 306 | (1955) |
| $^{102}$Pd | S 99.999 | 736-939 | S.S. | 56.75 | 9.56 | 213 | (1963) |
| $^{103}$Ru | S 99.99 | 793-945 | S.S. | 65.8 | 180 | 307 | (1959) |
| $^{35}$S | S 99.999 | 600-900 | R.A. | 40.0 | 1.65 | 308 | (1967) |

| Solute | Material | Temperature range (°C) | Form of analysis | Activation energy, Q (kcal/mole) | Frequency factor, $D_0$ (cm²/sec) | Reference No. | Year |
|---|---|---|---|---|---|---|---|
| **SILVER** | | | | | | | |
| $^{124}$Sb | S — | 650-950 | S.S. | 39.4 | 0.29 | 309 | (1952) |
| $^{124}$Sb | S,P 99.99 | 468-942 | S.S. | 38.32 | 0.169 | 310 | (1954) |
| $^{124}$Sb | P 99.999 | 780-950 *** | S.S.,R.A.*** | 39.07 | 0.234 | 311 | (1967) |
| $^{113}$Sn | S 99.99 | 592-937 | S.S. | 39.30 | 0.255 | 298 | (1954) |
| $^{204}$Tl | P — | 640-870 | S.S. | 37.9 | 0.15 | 302 | (1958) |
| $^{65}$Zn | S 99.99 | 640-925 | S.S. | 41.7 | 0.54 | 312 | (1955) |
| $^{65}$Zn | S | 785-895 | | 41.7 | 0.532 | 313 | (1967) |
| **SODIUM** | | | | | | | |
| $^{42}$K | | 0-91 | S.S. | 8.43 | 0.08 | 289 | (1967) |
| $^{86}$Rb | | 0-85 | S.S. | 8.49 | 0.15 | 289 | (1967) |
| **TANTALUM** | | | | | | | |
| $^{14}$C | P — | 600-2200 | — | 23.8 | $2.78 \times 10^{-3}$ | 314 | (1964) |
| $^{14}$C | P 99.01 | 1200-1600 | R.A. | 43.0 | $2.57 \times 10^{-2}$ | 261 | (1966) |
| $^{14}$C | | | | 40.3 | 0.012 | 315 | (1966) |
| $^{59}$Fe | P — | 930-1240 | — | 71.4 | 0.505 | 275 | (1955) |
| $^{95}$Nb | P,S 99.996 | 921-2484 | S.S. | 98.7 | 0.23 | 141 | (1965) |
| **TELLURIUM** | | | | | | | |
| $^{203}$Hg | P | 270-440 | — | 18.7 | $3.4 \times 10^{-5}$ | 316 | (1962) |
| $^{75}$Se | P | 320-440 | — | 28.6 | $2.6 \times 10^{-2}$ | 316 | (1962) |
| $^{204}$Tl | P | 360-430 | — | 41.0 | 320 | 317 | (1962) |
| $^{204}$Tl | S⊥c | 380-430 | — | 84.4 | $1.8 \times 10^{16}$ | 317 | (1962) |
| $^{204}$Tl | S∥c | 380-430 | — | 73.1 | $8.5 \times 10^{11}$ | 317 | (1962) |
| **α-THALLIUM** | | | | | | | |
| $^{110}$Ag | P⊥c 99.999 | 80-250 | S.S. | 11.8 | $3.8 \times 10^{-2}$ | 318 | (1968) |
| $^{110}$Ag | P∥c 99.999 | 80-250 | S.S. | 11.2 | $2.7 \times 10^{-2}$ | 318 | (1968) |
| $^{198}$Au | P⊥c 99.999 | 110-260 | S.S. | 2.8 | $2.0 \times 10^{-5}$ | 318 | (1968) |
| $^{198}$Au | P∥c 99.999 | 110-260 | S.S. | 5.2 | $5.3 \times 10^{-4}$ | 318 | (1968) |

| Solute | Material | Temperature range (°C) | Form of analysis | Activation energy, Q (kcal/mole) | Frequency factor, $D_0$ (cm²/sec) | Reference No. | Year |
|---|---|---|---|---|---|---|---|

**β-THALLIUM**

| Solute | Material | Temperature range (°C) | Form of analysis | Activation energy, Q (kcal/mole) | Frequency factor, $D_0$ (cm²/sec) | Reference No. | Year |
|---|---|---|---|---|---|---|---|
| $^{110}$Ag | P⊥c 99.999 | 230-310 | S.S. | 11.9 | $4.2 \times 10^{-2}$ | 318 | (1968) |
| $^{198}$Au | P∥c 99.999 | 230-310 | S.S. | 6.0 | $5.2 \times 10^{-4}$ | 318 | (1968) |

**α-THORIUM**

| $^{231}$Pa | P | 690-910 | | 71.6 | 395 | 143 | (1967) |
|---|---|---|---|---|---|---|---|
| $^{233}$U | P | 690-910 | | 79.3 | 2210 | 143 | (1967) |

**TIN**

| $^{110}$Ag | ∥c | 135-225 | S.S. | 12.3 | $7.1 \times 10^{-3}$ | 319 | (1966) |
|---|---|---|---|---|---|---|---|
| $^{110}$Ag | ⊥c | 135-225 | S.S. | 18.4 | 0.18 | 319 | (1966) |
| $^{110}$Ag | P | 478-500 | S.D. | 24.0 | 12.0 | 320 | (1967) |
| $^{198}$Au | ∥c | 135-225 | S.S. | 11.0 | $5.8 \times 10^{-3}$ | 319 | (1966) |
| $^{198}$Au | ⊥c | 135-225 | S.S. | 17.7 | 0.16 | 319 | (1966) |
| $^{60}$Co | S,P | 140-217 | R.A. | 22.0 | 5.5 | 321 | (1960) |
| $^{114}$In | S⊥c 99.998 | 181-221 | S.S. | 25.8 | 34.1 | 322 | (1958) |
| $^{114}$In | S∥c 99.998 | 181-221 | S.S. | 25.6 | 12.2 | 322 | (1958) |
| $^{65}$Zn | S,P | 140-217 | R.A. | 7.8 | $9.8 \times 10^{-4}$ | 321 | (1960) |

**β-TITANIUM**

| $^{14}$C | P — | 1150-1650 | — | 19.8 | $3.18 \times 10^{-3}$ | 323 | (1963) |
|---|---|---|---|---|---|---|---|
| $^{14}$C | P 99.62 | 1100-1400 | R.A. | 20.0 | $3.02 \times 10^{-3}$ | 261 | (1960) |
| $^{51}$Cr | P — | 500-1300 | A.R.G. | 67 | 200 | 175 | (1958) |
| $^{51}$Cr | P 99.74 | 1000-1200 | A.R.G. | 37.7 | $1 \times 10^{-2}$ | 324 | (1959) |
| $^{51}$Cr | P 99.24 | 1000-1200 | A.R.G. | 35.3 | $5 \times 10^{-3}$ | 324 | (1959) |
| $^{51}$Cr | P 99.7 | 950-1600 | A.R.G. *** | 35.1, *** 61 | $5 \times 10^{-3}$ 4.9 | 325 | (1963) |
| $^{60}$Co | P 99.7 | 900-1250 | S.S. | 30.6 | $1.2 \times 10^{-2}$ | 326 | (1962) |
| $^{60}$Co | P 99.7 | 900-1600 | S.S. *** | 30.6, *** 52.5 | $1.2 \times 10^{-2}$, 2.0 | 325 | (1963) |
| $^{55}$Fe | P — | 500-1300 | A.R.G. | 50 | 20 | 175 | (1958) |
| $^{55}$Fe | P 99.7 | 900-1250 | A.R.G. | 31.6 | $7.8 \times 10^{-3}$ | 326 | (1962) |
| $^{59}$Fe | P 99.7 | 900-1250 | A.R.G. | 32.9 | $1.23 \times 10^{-2}$ | 326 | (1962) |
| $^{55}$Fe | — | — | A.R.G. | 32.0 | $8.5 \times 10^{-3}$ | 326 | (1962) |
| $^{55}$Fe | P 99.7 | 900-1600 | A.R.G. *** | 31.6, *** 55 | $7.8 \times 10^{-3}$, 2.7 | 325 | (1963) |
| $^{99}$Mo | P 99.7 | 1000-1600 | S.S. *** | 43.0, *** 73 | $8 \times 10^{-3}$, 20 | 325 | (1963) |

| Solute | Material | Temperature range (°C) | Form of analysis | Activation energy, Q (kcal/mole) | Frequency factor, $D_0$ (cm²/sec) | Reference No. | Year |
|---|---|---|---|---|---|---|---|
| β -TITANIUM (continued) | | | | | | | |
| $^{99}$Mo | P 98.94 | 900-1100 | | 33.2 | $2.82 \times 10^{-4}$ | 327 | (1967) |
| $^{99}$Mo | P 98.94 | 1100-1560 | | 51.3 | 0.24 | 327 | (1967) |
| $^{54}$Mn | P 99.7 | 900-1250 | S.S. | 33.7 | $6.1 \times 10^{-3}$ | 326 | (1962) |
| $^{54}$Mn | P 99.7 | 900-1600 | S.S.*** | 33.7,*** 58 | $6.1 \cdot 10^{-3}$, 20 | 325 | (1963) |
| $^{95}$Nb | P 99.7 | 1000-1250 | A.R.G. | 39.3 | $5.0 \times 10^{-3}$ | 326 | (1962) |
| $^{95}$Nb | P 99.7 | 1000-1600 | A.R.G. *** | 39.3,*** 73 | $5.0 \times 10^{-3}$, 20 | 325 | (1963) |
| $^{63}$Ni | P 99.7 | 900-1250 | A.R.G. | 29.6 | $9.2 \times 10^{-3}$ | 326 | (1962) |
| $^{63}$Ni | P 99.7 | 295-1600 | A.R.G.*** | 29.6,*** 52.5 | $9.2 \times 10^{-3}$, 2.0 | 325 | (1963) |
| $^{32}$P | P 99.7 | 950-1600 | S.S. | 24.1, 56.5 | $3.62 \times 10^{-3}$, 5 | 328 | (1965) |
| $^{46}$Sc | P 99.7 | 919-1290 | S.S. | 32.7 | $2.1 \times 10^{-3}$ | 328 | (1965) |
| $^{113}$Sn | P 99.78 | 900-1100 | S.D. | 78.0 | 1630 | 329 | (1960) |
| $^{113}$Sn | P 99.3 | 900-1100 | S.D. | 86.5 | 10 | 329 | (1960) |
| $^{113}$Sn | P 99.7 | 950-1600 | S.S. *** | 31.6,*** 69.2 | $3.8 \times 10^{-4}$, 10 | 328 | (1965) |
| $^{235}$U | P 99.62 | 915-1200 | R.A. | 29.3 | $4.9 \times 10^{-3}$ | 267 | (1966) |
| $^{48}$V | P.,99.95 | 900-1545 | S.S. | 32.2, 57.2 | $3.1 \times 10^{-4}$, 1.4 | 330 | (1964) |
| $^{48}$V | P 99.99 | 1100-1800 | S.S. | Nonlinear | | 331 | (1968) |
| $^{185}$W | P — | 500-1300 | A.R.G. | 49 | 0.3 | 175 | (1958) |
| $^{185}$W | P 98.94 | 900-1250 | | *** 43.9 *** | $3.6 \times 10^{-3}$ | 327 | (1967) |
| $^{95}$Zr | P 98.94 | 920-1500 | | 35.4 | $4.7 \times 10^{-3}$ | 327 | (1967) |
| TUNGSTEN | | | | | | | |
| $^{14}$C | P 99.51 | 1200-1600 | R.A. | 53.5 | $8.9 \times 10^{-3}$ | 261 | (1966) |
| $^{59}$Fe | P — | 940-1240 | — | 66.0 | $1.4 \times 10^{-2}$ | 275 | (1955) |
| $^{99}$Mo | S | 2000-2400 | S.S. | 121 | $5 \times 10^{-2}$ | 332 | (1967) |
| $^{99}$Mo | P | 2100-2400 | S.S. | 110 | $3.7 \times 10^{-3}$ | 332 | (1967) |
| $^{186}$Re | S 99.92 | 2660-3230 | S.S. | 162.8 | 275 | 156 | (1965) |
| $^{186}$Re | P | 2100-2400 | S.S. *** | 141 *** | 19.5 | 332 | (1967) |
| γ -URANIUM | | | | | | | |
| $^{195}$Au | P 99.99 | 785-1007 | S.S. | 30.4 | $4.86 \times 10^{-3}$ | 333 | (1961) |
| $^{60}$Co | P 99.99 | 783-989 | S.S. | 12.57 | $3.51 \times 10^{-4}$ | 334 | (1964) |
| $^{51}$Cr | P — | 844-948 | S.S. | 21.8 | $1.98 \times 10^{-3}$ | 335 | (1961) |

| Solute | Material | Temperature range (°C) | Form of analysis | Activation energy, Q (kcal/mole) | Frequency factor, $D_0$ (cm²/sec) | Reference No. | Year |
|---|---|---|---|---|---|---|---|

γ -URANIUM

| Solute | Material | Temperature range (°C) | Form of analysis | Activation energy, Q (kcal/mole) | Frequency factor, $D_0$ (cm²/sec) | Reference No. | Year |
|---|---|---|---|---|---|---|---|
| $^{51}$Cr | P 99.99 | 797-1037 | S.S.*** | 24.46*** | $5.47 \times 10^{-3}$ | 334 | (1964) |
| $^{64}$Cu | P 99.99 | 787-1039 | S.S. | 24.06 | $1.96 \times 10^{-3}$ | 334 | (1964) |
| $^{55}$Fe | P 99.99 | 787-990 | S.S. | 12.01 | $2.69 \times 10^{-4}$ | 334 | (1964) |
| $^{54}$Mn | P 99.99 | 787-939 | S.S. | 13.88 | $1.81 \times 10^{-4}$ | 334 | (1964) |
| $^{63}$Ni | P 99.99 | 787-1039 | S.S. | 15.66 | $5.36 \times 10^{-4}$ | 334 | (1964) |
| $^{95}$Nb | P 99.99 | 791-1102 | S.S. | 39.65 | $4.87 \times 10^{-2}$ | 334 | (1964) |

VANADIUM

| Solute | Material | Temperature range (°C) | Form of analysis | Activation energy, Q (kcal/mole) | Frequency factor, $D_0$ (cm²/sec) | Reference No. | Year |
|---|---|---|---|---|---|---|---|
| $^{51}$Cr | P 99.8 | 960-1200 | R.A. | 64.6 | $9.54 \times 10^{-3}$ | 195 | (1962) |
| $^{55}$Fe | S,P 99.97 | 842-1171 | S.S. | 70.5 | 0.60 | 164 | (1965) |
| $^{44}$Ti | P 99.98 | 1100-1800 | S.S. | | Nonlinear | 331 | (1968) |

ZINC

| Solute | Material | Temperature range (°C) | Form of analysis | Activation energy, Q (kcal/mole) | Frequency factor, $D_0$ (cm²/sec) | Reference No. | Year |
|---|---|---|---|---|---|---|---|
| $^{110}$Ag | S⊥c 99.999 | 271-413 | S.S. | 27.6 | 0.45 | 336 | (1961) |
| $^{110}$Ag | S∥c 99.999 | 271-413 | S.S. | 26.0 | 0.32 | 336 | (1961) |
| $^{198}$Au | S⊥c 99.999 | 315-415 | S.S. | 29.72 | 0.29 | 337 | (1963) |
| $^{198}$Au | S∥c 99.999 | 315-415 | S.S. | 29.73 | 0.97 | 337 | (1963) |
| $^{115}$Cd | S⊥c 99.999 | 225-416 | S.S. | 20.12 | 0.117 | 337 | (1963) |
| $^{115}$Cd | S∥c 99.999 | 225-416 | S.S. | 20.54 | 0.114 | 337 | (1963) |
| $^{64}$Cu | S∥c 99.999 | 338-415 | S.S. | 29.53 | 2.22 | 338 | (1966) |
| $^{64}$Cu | S⊥c 99.999 | 338-415 | S.S. | 29.92 | 2.0 | 338 | (1966) |
| $^{72}$Ga | S∥c 99.999 | 240-403 | S.S. | 18.4 | 0.016 | 338 | (1966) |
| $^{72}$Ga | S⊥c 99.999 | 240-403 | S.S. | 18.15 | 0.018 | 338 | (1966) |
| $^{203}$Hg | P 99.96 | 20-200 | — | 3.2 | $5.8 \times 10^{-10}$ | 339 | (1963) |
| $^{203}$Hg | S⊥c 99.999 | 260-413 | S.S. | 20.18 | 0.073 | 340 | (1967) |
| $^{203}$Hg | S∥c 99.999 | 260-413 | S.S. | 19.70 | 0.056 | 340 | (1967) |
| $^{114}$In | S⊥c 99.999 | 271-413 | S.S. | 19.6 | 0.14 | 336 | (1961) |
| $^{114}$In | S∥c 99.999 | 271-413 | S.S. | 19.1 | 0.062 | 336 | (1961) |

α-ZIRCONIUM

| Solute | Material | Temperature range (°C) | Form of analysis | Activation energy, Q (kcal/mole) | Frequency factor, $D_0$ (cm²/sec) | Reference No. | Year |
|---|---|---|---|---|---|---|---|
| $^{51}$Cr | P 99.9 | 700-850 | R.A. | 18.0 | $1.19 \times 10^{-8}$ | 341 | (1965) |
| $^{55}$Fe | P  — | 750-840 | — | 48.0 | $2.5 \times 10^{-2}$ | 342 | (1964) |
| $^{99}$Mo | | | R.A. | 24.7 | $6.2 \times 10^{-8}$ | 343 | (1967) |
| $^{113}$Sn | P  — | 300-700 | A.R.G. | 22.0 | $1 \times 10^{-8}$ | 175 | (1958) |

| Solute | Material | Temperature range (°C) | Form of analysis | Activation energy, Q (kcal/mole) | Frequency factor, $D_0$ (cm²/sec) | Reference No. | Year |
|---|---|---|---|---|---|---|---|
| α-ZIRCONIUM | | | | | | | |
| $^{182}$Ta | P 99.6 | 700-800 | R.A. | 70.0 | 100 | 344 | (1960) |
| $^{48}$V | P 99.99 | 600-850 | R.A. | 22.9 | $1.12 \times 10^{-8}$ | 165 | (1968) |
| β-ZIRCONIUM | | | | | | | |
| $^{14}$C | P 99.9 | 900-1260 | | 26.7 | $4.8 \times 10^{-3}$ | 346 | (1965) |
| $^{14}$C | P 99.6 | 1100-1600 | S.S. | 34.2 | $3.57 \times 10^{-2}$ | 346 | (1966) |
| $^{51}$Cr | P 99.9 | 880-1100 | R.A. | 41.0 | $3.85 \times 10^{-2}$ | 341 | (1964) |
| $^{51}$Cr | P 99.7 | 900-1200 | R.A.*** | 32.0*** | $4.17 \times 10^{-3}$ | 347 | (1967) |
| $^{55}$Fe | P — | 890-1100 | — | 30.0 | $4 \times 10^{-2}$ | 342 | (1964) |
| $^{59}$Fe | P 99.7 | 900-1400 | R.A.*** | 27.0*** | $9.1 \times 10^{-3}$ | 347 | (1967) |
| $^{99}$Mo | P 99.7 | 900-1200 | R.A.*** | 44.4*** | $3.63 \times 10^{-2}$ | 347 | (1967) |
| $^{99}$Mo | P 99.7 | 1355-1560 | R.A.*** | 58.2*** | . 1.29 | 347 | (1967) |
| $^{99}$Mo | P | | R.A. | 68.55, 35.2 | 2.63, $1.99 \times 10^{-4}$ | 343 | (1967) |
| $^{95}$Nb | P 99.9 | 882-1758 | S.S. | 25.1 + 35.5 × (T−1136) | $9 \times 10^{-6} \times$ (T/1136)$^{18.1}$ | 348 | (1963) |
| $^{113}$Sn | P — | 1000-1250 | R.A. | 39.0 | $5 \times 10^{-3}$ | 178 | (1959) |
| $^{182}$Ta | P 99.6 | 900-1200 | R.A. | 27.0 | $5.5 \times 10^{-5}$ | 244 | (1960) |
| $^{48}$V | P 99.99 | 870-1200 | R.A. | 45.8 | $7.59 \times 10^{-3}$ | 347 | (1967) |
| $^{48}$V | P 99.99 | 1200-1400 | R.A. | 57.2 | 0.32 | 347 | (1967) |
| $^{185}$W | P 99.7 | 900-1250 | R.A. | 55.8 | 0.41 | 347 | (1967) |

# Part III

## Self- and Impurity Diffusion in Alloys

| Solute | Composition | Temperature range (°C) | Form of analysis | Activation energy, Q (kcal/mole) | Frequency factor, $D_0$ (cm²/sec) | No. | Year |
|---|---|---|---|---|---|---|---|

ALUMINUM

| Solute | Composition | Temperature range (°C) | Form of analysis | Activation energy, Q (kcal/mole) | Frequency factor, $D_0$ (cm²/sec) | No. | Year |
|---|---|---|---|---|---|---|---|
| $^{110}$Ag | 30 wt% Ag | 400-595 | | 28.9 | 0.39 | 349 | (1968) |
| $^{60}$Co | 44.5 at% Ni | 1050-1350 | R.A. | 47.1 | $7.2 \times 10^{-3}$ | 350 | (1954) |
| $^{60}$Co | 46.9 at% Ni | 1050-1350 | R.A. | 67.6 | 2.6 | 350 | (1954) |
| $^{60}$Co | 49.3 at% Ni | 1050-1350 | R.A. | 80.6 | 57.7 | 350 | (1954) |
| $^{65}$Zn | 4.33 at% Zn | 360-610 | S.S. | 28.3 | 0.35 | 173 | (1959) |
| $^{65}$Zn | 9.23 at% Zn | 360-575 | S.S. | 27.0 | 0.2 | 173 | (1959) |
| $^{65}$Zn | 16.7 at% Zn | 360-525 | S.S. | 25.0 | 0.1 | 173 | (1959) |
| $^{65}$Zn | 36.9 at% Zn | 360-450 | S.S. | 24.1 | 0.16 | 173 | (1959) |
| $^{65}$Zn | 49.4 at% Zn | 325-440 | S.S. | 21.9 | 0.048 | 173 | (1959) |

CHROMIUM

| Solute | Composition | Temperature range (°C) | Form of analysis | Activation energy, Q (kcal/mole) | Frequency factor, $D_0$ (cm²/sec) | No. | Year |
|---|---|---|---|---|---|---|---|
| $^{51}$Cr | 16 at% Fe | 1040-1250 | | 64.2 | 0.376 | 351 | (1958) |
| $^{51}$Cr | 31 at% Fe | 1000-1275 | R.A.,S.D. | 75.5 | 24.6 | 352 | (1960) |
| $^{51}$Cr | 49 at% Fe | 1000-1275 | R.A. | 70.0 | 40 | 352 | (1960) |
| $^{59}$Fe | 9.4 at% Fe | 1000-1250 | R.A. | 52.3 | $5 \times 10^{-4}$ | 353 | (1965) |
| $^{59}$Fe | 16 at% Fe | 1040-1250 | | 81.9 | 146 | 351 | (1958) |
| $^{59}$Fe | 19 at% Fe | 957-1250 | S.S. | 103.2 | $3.4 \times 10^{5}$ | 354 | (1958) |
| $^{59}$Fe | 19.0 at% Fe | 1000-1150 | R.A. | 68.0 | 1.0 | 353 | (1965) |
| $^{59}$Fe | 28.5 at% Fe | 950-1250 | R.A. | 74.5 | 32.0 | 353 | (1965) |
| $^{59}$Fe | 38.5 at% Fe | 957-1250 | S.S. | 88.2 | $10^{4}$ | 354 | (1958) |
| $^{59}$Fe | 38.5 at% Fe | 1050-1150 | R.A. | 80.0 | 200 | 353 | (1965) |

| Solute | Composition | Temperature range (°C) | Form of analysis | Activation energy, Q (kcal/mole) | Frequency factor, $D_0$ (cm²/sec) | Reference No. | Year |
|---|---|---|---|---|---|---|---|

CHROMIUM  (continued)

| Solute | Composition | Temperature range (°C) | Form of analysis | Activation energy, Q (kcal/mole) | Frequency factor, $D_0$ (cm²/sec) | Reference No. | Year |
|---|---|---|---|---|---|---|---|
| $^{59}$Fe | 43.5 at% Fe | 957-1250 | S.S. | 81.4 | 1500 | 354 | (1958) |
| $^{59}$Fe | 47.2 at% Fe | 957-1250 | S.S. | 84.7 | 7700 | 354 | (1958) |
| $^{59}$Fe | 48.3 at% Fe | 1000-1200 | R.A. | 68.8 | 27.0 | 353 | (1965) |
| $^{59}$Fe | 49.0 at% Fe | 1040-1250 | | 74.6 | 250 | 351 | (1958) |

COBALT

| Solute | Composition | Temperature range (°C) | Form of analysis | Activation energy, Q (kcal/mole) | Frequency factor, $D_0$ (cm²/sec) | Reference No. | Year |
|---|---|---|---|---|---|---|---|
| $^{60}$Co | 10% Al | 1040-1250 | R.A. | 67.5 | 2.65 | 355 | (1956) |
| $^{60}$Co | 42% Al | 1000-1250 | R.A. | 85 | 184 | 356 | (1956) |
| $^{60}$Co | 49% Al | 900-1200 | R.A. | 102 | $3.3 \times 10^4$ | 356 | (1956) |
| $^{60}$Co | 50% Al | 900-1100 | R.A. | 105 | | 356 | (1956) |
| $^{60}$Co | 4 at% Cr | 1100-1350 | | 65.8 | 0.67 | 357 | (1955) |
| $^{60}$Co | 6.75% Cr | 1020-1240 | R.A. | 73.3 | | 355 | (1956) |
| $^{60}$Co | 7 at% Cr | 1100-1350 | | 79.3 | 56.3 | 357 | (1955) |
| $^{60}$Co | 11.5% Cr | 1020-1240 | R.A. | 86.3 | | 355 | (1956) |
| $^{60}$Co | 16.5% Cr | 1020-1240 | R.A. | 102 | | 355 | (1956) |
| $^{60}$Co | 27.5% Cr | 1020-1240 | R.A. | 110 | | 355 | (1956) |
| $^{60}$Co | 15 at% Fe 4 at% Ti | | | 51.2 | $8 \times 10^{-3}$ | 357 | (1956) |
| $^{60}$Co | 21 at% Fe | | | 65.0 | 0.54 | 357 | (1955) |
| $^{60}$Co | 50 at% Fe | 840-925 | S.S. | 27.4 | $2.6 \times 10^{-6}$ | 358 | (1958) |
| $^{60}$Co | 50% Fe | 1050-1200 | R.A. | 37.0 | $5 \times 10^{-6}$ | 355 | (1956) |
| $^{60}$Co | 50 at% Fe | 1000-1250 | S.S. | 41.8 | $1.1 \times 10^{-4}$ | 358 | (1958) |
| $^{60}$Co | 4 at% Ni | | | 87.4 | 838 | 357 | (1955) |
| $^{60}$Co | 6.2 at% Ni | 1213-1368 | R.A. | 72.9 | 7.4 | 25 | (1965) |
| $^{60}$Co | 8 at% Ni | | | 81.2 | 124 | 325 | (1955) |
| $^{60}$Co | 10.9 at% Ni | 864-1048 | | 67.1 | 0.61 | 24 | (1962) |
| $^{60}$Co | 10.9 at% Ni | 1144-1297 | R.A. | 63.6 | 0.21 | 24 | (1962) |
| $^{60}$Co | 11.0 at% Ni | 1163-1393 | R.A. | 69.3 | 2.52 | 25 | (1965) |
| $^{60}$Co | 19.4 at% Ni | 1163-1393 | R.A. | 65.9 | 0.99 | 25 | (1965) |
| $^{60}$Co | 19.5% Ni 20.6% Mn | 1000-1200 | R.A. | 54.5 | 0.05 | 355 | (1956) |
| $^{60}$Co | 20.1 at% Ni | 845-988 | R.A. | 73.5 | 5.96 | 24 | (1962) |
| $^{60}$Co | 20.1 at% Ni | 1090-1246 | R.A. | 71.0 | 2.42 | 24 | (1962) |
| $^{60}$Co | 26.0 at% Ni 9 at% Cr | 1100-1350 | | 72.1 | 6.3 | 357 | (1955) |
| $^{60}$Co | 26.0 at% Ni 18 at% Cr | 1100-1350 | | 64.2 | 0.4 | 357 | (1955) |

| Solute | Composition | Temper- ature range (°C) | Form of analy- sis | Activation energy, Q (kcal/mole) | Frequency factor, $D_0$ (cm²/sec) | Reference No. | Year |
|---|---|---|---|---|---|---|---|
| COBALT | (continued) | | | | | | |
| [60]Co | 27.4 at%Ni | 1163-1393 | R.A. | 64.6 | 0.70 | 25 | (1965) |
| [60]Co | 30 at%Ni | | | 60.3 | 0.155 | 357 | (1955) |
| [60]Co | 30.1 at%Ni 0.26 at%Si | 772-899 | R.A. | 68.6 | 1.16 | 24 | (1962) |
| [60]Co | 30.1 at%Ni 0.26 at%Si | 1048-1246 | R.A. | 67.0 | 0.78 | 24 | (1962) |
| [60]Co | 0.26 at%Si | 778-1048 | R.A. | 65.4 | 0.5 | 24 | (1962) |
| [60]Co | 0.26 at%Si | 1192-1297 | R.A. | 62.2 | 0.17 | 24 | (1962) |
| [63]Ni | 6.2 at%Ni | 1213-1368 | R.A. | 72.2 | 5.4 | 25 | (1965) |
| [63]Ni | 10.9 at%Ni | 864-1048 | R.A. | 65.6 | 0.46 | 24 | (1962) |
| [63]Ni | 10.9 at%Ni | 1144-1297 | R.A. | 62.7 | 0.17 | 24 | (1962) |
| [63]Ni | 11.0 at%Ni | 1241-1410 | R.A. | 72.6 | 6.42 | 25 | (1965) |
| [63]Ni | 19.4 at%Ni | 1163-1410 | R.A. | 68.2 | 1.89 | 25 | (1965) |
| [63]Ni | 20.1 at%Ni | 864-988 | R.A. | 69.7 | 1.66 | 24 | (1962) |
| [63]Ni | 20.1 at%Ni | 1060-1246 | R.A. | 65.7 | 0.41 | 24 | (1962) |
| [63]Ni | 27.4 at%Ni | 1163-1410 | R.A. | 67.2 | 1.60 | 25 | (1965) |
| [63]Ni | 30.1 at%Ni | 772-899 | R.A. | 68.5 | 2.01 | 24 | (1962) |
| [63]Ni | 30.1 at%Ni | 1048-1246 | R.A. | 64.8 | 0.67 | 24 | (1962) |
| [63]Ni | 0.26 at%Si | 772-1048 | R.A. | 64.3 | 0.34 | 24 | (1962) |
| [63]Ni | 0.26at%Si | 1192-1297 | R.A. | 60.2 | 0.10 | 24 | (1962) |
| [185]W | 0.82 wt%W 0.14 wt%Fe | 1150-1350 | | 68.0 | 2.88 | 359 | (1961) |
| COPPER | | | | | | | |
| [64]Cu | 1 wt%Ni | 780-890 | | 48.9 | 1.86 | 33 | (1967) |
| [64]Cu | 21.5 at%Ni | 863-1112 | S.S. | 55.3 | 1.9 | 32 | (1964) |
| [64]Cu | 9.8 at%Pt | 899-1046 | S.S. | 52.8 | 1.1 | 214 | (1963) |
| [64]Cu | 24.6 at%Pt | 947-1096 | S.S. | 54.7 | 0.53 | 214 | (1963) |
| [64]Cu | 49.4 at%Pt | 1000-1293 | S.S. | 51.0 | 0.027 | 214 | (1963) |
| [64]Cu | 18.0 at%Sn | 600-725 | S.S. | 29.2 | 0.32 | 360 | (1968) |
| [64]Cu | 19.8 at%Sn | 600-725 | S.S. | 27.1 | $9.2 \times 10^{-2}$ | 360 | (1968) |
| [64]Cu | 20.5 at%Sn | 440-575 | S.S. | 30.9 | 4.7 | 360 | (1968) |
| [64]Cu | 27 wt%Zn | 800-900 | S.S. | 44.5 | 0.85 | 361 | (1954) |
| [64]Cu | 31 wt%Zn | 600-900 | S.S. | 41.9 | 0.34 | 218 | (1957) |
| [64]Cu | 45 wt%Zn | 640-870 | S.S. | 25.0 | 0.038 | 361 | (1954) |
| [64]Cu | 48.0 at%Zn | 497-817 | S.S. | 22.0 | 0.011 | 362 | (1956) |
| [63]Ni | 21.5 at%Ni | 930-1113 | S.S. | 49.7 | 0.063 | 32 | (1964) |

| Solute | Composition | Temperature range (°C) | Form of analysis | Activation energy, Q (kcal/mole) | Frequency factor, $D_0$ (cm²/sec) | Reference No. | Year |
|---|---|---|---|---|---|---|---|

COPPER (continued)

| Solute | Composition | Temperature range (°C) | Form of analysis | Activation energy, Q (kcal/mole) | Frequency factor, $D_0$ (cm²/sec) | Reference No. | Year |
|---|---|---|---|---|---|---|---|
| $^{195}$Pt | 9.8 at% Pt | 906-1058 | S.S. | 52.6 | 0.093 | 214 | (1963) |
| $^{195}$Pt | 24.6 at% Pt | 946-1094 | S.S. | 51.4 | 0.019 | 214 | (1963) |
| $^{195}$Pt | 49.4 at% Pt | 1034-1287 | S.S. | 59.5 | 0.066 | 214 | (1963) |
| $^{124}$Sb | 48.0 at% Zn | 351-594 | S.S. | 23.5 | 0.08 | 315 | (1956) |
| $^{113}$Sn | 18.0 at% Sn | 600-725 | S.S. | 17.8 | $1.4 \times 10^{-2}$ | 360 | (1968) |
| $^{113}$Sn | 19.8 at% Sn | 600-725 | S.S. | 10.2 | $3.6 \times 10^{-3}$ | 360 | (1968) |
| $^{113}$Sn | 20.5 at% Sn | 440-575 | S.S. | 49.7 | 2400 | 360 | (1968) |
| $^{65}$Zn | 10.0 at% Ni | 800-900 | — | 44 | — | 219 | (1966) |
| $^{65}$Zn | 10.0 at% Ni +10.0 at% Zn | 800-900 | — | 47 | — | 219 | (1966) |
| $^{65}$Zn | 10.0 at% Ni +20.0 at% Zn | 800-900 | | 41 | — | 219 | (1966) |
| $^{65}$Zn | 10.0 at% Ni +30.0 at% Zn | 800-900 | | 34 | — | 219 | (1966) |
| $^{65}$Zn | 20.0 at% Ni | 800-900 | | 54 | — | 219 | (1966) |
| $^{65}$Zn | 20.0 at% Ni +10.0 at% Zn | 800-900 | | 45 | — | 219 | (1966) |
| $^{65}$Zn | 20.0 at% Ni +20.0 at% Zn | 800-900 | | 53 | — | 219 | (1966) |
| $^{65}$Zn | 20.0 at% Ni +30.0 at% Zn | 800-900 | | 42 | — | 219 | (1966) |
| $^{65}$Zn | 10 at% Zn | 800-900 | | 47 | — | 219 | (1966) |
| $^{65}$Zn | 20 at% Zn | 800-900 | | 38 | — | 219 | (1966) |
| $^{65}$Zn | 27.0 wt% Zn | 800-900 | S.S. | 41.16 | 0.85 | 361 | (1954) |
| $^{65}$Zn | 30.0 at% Zn | 800-900 | | 45 | — | 219 | (1966) |
| $^{65}$Zn | 31.0 wt% Zn | 600-900 | S.S. | 40.07 | 0.73 | 218 | (1957) |
| $^{65}$Zn | 45.0 wt% Zn | 640-870 | S.S. | 23.3 | 0.031 | 361 | (1954) |
| $^{65}$Zn | 48.0 at% Zn | 499-718 | S.S. | 18.78 | $3.5 \times 10^{-3}$ | 362 | (1956) |

GOLD

| Solute | Composition | Temperature range (°C) | Form of analysis | Activation energy, Q (kcal/mole) | Frequency factor, $D_0$ (cm²/sec) | Reference No. | Year |
|---|---|---|---|---|---|---|---|
| $^{110}$Ag | 6 at% Ag | 660-961 | S.S. | 40.26 | 0.72 | 363 | (1963) |
| $^{110}$Ag | 17 at% Ag | 650-1010 | S.S. | 41.02 | 0.09 | 363 | (1963) |
| $^{110}$Ag | 34 at% Ag | 654-971 | S.S. | 41.73 | 0.11 | 363 | (1963) |
| $^{198}$Au | 6 at% Ag | 718-1010 | S.S. | 42.08 | 0.09 | 363 | (1963) |
| $^{198}$Au | 17 at% Ag | 712-1001 | S.S. | 43.05 | 0.12 | 363 | (1963) |
| $^{198}$Au | 17.1 wt% Ag | 717-992 | S.S. | 40.7 | 0.041 | 42 | (1957) |
| $^{198}$Au | 34 at% Ag | 715-1001 | S.S. | 44.51 | 0.17 | 363 | (1963) |

| Solute | Composition | Temperature range (°C) | Form of analysis | Activation energy, Q (kcal/mole) | Frequency factor, $D_0$ (cm²/sec) | Reference No. | Year |
|---|---|---|---|---|---|---|---|
| **GOLD** (continued) | | | | | | | |
| $^{198}$Au | 50 at% Cd | 303-604 | S.S. | 27.9 | 0.17 | 364 | (1961) |
| $^{198}$Au | 47.5% Cd | 354-600 | S.S. | 28.1 | 0.23 | 365 | (1967) |
| $^{198}$Au | 49.0% Cd | 440-550 | S.S. | 30.0 | 0.61 | 365 | (1967) |
| $^{198}$Au | 50.0% Cd | 440-550 | S.S. | 27.9 | 0.17 | 365 | (1967) |
| $^{198}$Au | 50.5% Cd | 440-550 | S.S. | 26.2 | 0.12 | 365 | (1967) |
| $^{115}$Cd | 50 at% Cd | 330-620 | S.S. | 28.0 | 0.23 | 364 | (1961) |
| $^{115}$Cd | 47.5% Cd | 354-600 | S.S. | 31.0 | 1.36 | 365 | (1967) |
| $^{115}$Cd | 49.0% Cd | 440-550 | S.S. | 31.2 | 1.50 | 365 | (1967) |
| $^{115}$Cd | 50.0% Cd | 440-550 | S.S. | 28.0 | 0.23 | 365 | (1967) |
| $^{115}$Cd | 50.5% Cd | 440-550 | S.S. | 27.1 | 0.22 | 365 | (1967) |
| $^{63}$Ni | 10 at% Ni | 880-940 | S.S. | 47.6 | 0.8 | 83 | (1957) |
| $^{63}$Ni | 20 at% Ni | 880-940 | S.S. | 47.8 | 0.82 | 83 | (1957) |
| $^{63}$Ni | 36 at% Ni | 873-920 | S.S. | 49.1 | 1.10 | 83 | (1957) |
| **HAFNIUM** | | | | | | | |
| $^{14}$C | 1.5 wt% Zr 0.15 wt% Fe | 1120-1760 | | 74.6 | 74 | 366 | (1968) |
| $^{14}$C | 1.5 wt% Zr 0.15 wt% Fe | 1820-2130 | | 40.0 | $4.2 \times 10^{-2}$ | 366 | (1968) |
| $^{181}$Hf | 2.7 wt% Zr | 1785-2158 | | 43.8 | $4.8 \times 10^{-3}$ | 366 | (1968) |
| **IRON** | | | | | | | |
| $^{14}$C | 0.92 wt% Cr | 500-800 | R.A. | 33.6 | 16.4 | 234 | (1960) |
| $^{14}$C | 0.56 wt% Mo | 500-800 | R.A. | 28.9 | 2.0 | 234 | (1960) |
| $^{14}$C | 2.58 wt% Mo | 500-800 | R.A. | 33.5 | 20.0 | 234 | (1960) |
| $^{14}$C | 0.46 wt% Ni | 500-800 | R.A. | 24.8 | 0.2 | 234 | (1960) |
| $^{14}$C | 2.0 wt% Ni | 500-800 | R.A. | 26.6 | 0.3 | 234 | (1960) |
| $^{14}$C | 0.79 wt% Si | 500-800 | R.A. | 26.0 | 0.4 | 234 | (1960) |
| $^{14}$C | 2.38 wt% Si | 500-800 | R.A. | 27.2 | 0.8 | 234 | (1960) |
| $^{14}$C | 2.5 wt% Si 0.8 wt% C | 500-800 | R.A. | 34.5 | 0.12 | 234 | (1960) |
| $^{14}$C | 3.6 wt% Si | 500-800 | R.A. | 29.3 | 2.2 | 234 | (1960) |
| $^{60}$Co | 50 at% Co | 1000-1250 | R.A. | 41.8 | $1.1 \times 10^{-4}$ | 358 | (1958) |
| $^{60}$Co | 50 at% Co | 840-925 | R.A. | 27.4 | $2.6 \times 10^{-6}$ | 358 | (1958) |
| $^{60}$Co | 0.8 wt% C 0.4 wt% Mn | 1050-1250 | R.A. | 80.0 | 90 | 237 | (1954) |

| Solute | Composition | Temper-ature range (°C) | Form of analy-sis | Activation energy, Q (kcal/mole) | Frequency factor, $D_0$ (cm²/sec) | Reference No. | Year |
|---|---|---|---|---|---|---|---|

IRON (continued)

| Solute | Composition | Temper-ature range (°C) | Form of analy-sis | Activation energy, Q (kcal/mole) | Frequency factor, $D_0$ (cm²/sec) | Reference No. | Year |
|---|---|---|---|---|---|---|---|
| [60]Co | 18 wt% Cr<br>17 wt% Ni<br>5 wt% Al | 900-1200 | S.D. | 68.0 | 0.37 | 367 | (1960) |
| [60]Co | 15 wt% Cr<br>18 wt% Ni<br>5 wt% Al,<br>1 wt% Mo | 900-1200 | S.D. | 78.0 | 58 | 367 | (1960) |
| [60]Co | 15 wt% Cr<br>18 wt% Ni<br>5 wt% Al,<br>1 wt% Nb | 900-1200 | S.D. | 71.1 | 111 | 367 | (1960) |
| [60]Co | 15 wt% Cr<br>18 wt% Ni<br>5 wt% Al,<br>1 wt% Zr | 900-1200 | S.D. | 74.5 | 16.4 | 367 | (1960) |
| [60]Co | 15 wt% Cr<br>18 wt% Ni<br>5 wt% Al,<br>1/2 wt% Nb,<br>1/2 wt% Zr | 900-1200 | S.D. | 76.5 | 206 | 367 | (1960) |
| [60]Co | 18 wt% Cr<br>18 wt% Ni<br>8 wt% Al,<br>2 wt% Mo | 800-1000 | S.D. | 66.4 | 1.26 | 367 | (1960) |
| [60]Co | 18 wt% Cr<br>17 wt% Ni<br>8 wt% Al,<br>1 wt% Mo,<br>1/2 wt% Nb | 800-1000 | S.D. | 59.8 | 103 | 367 | (1960) |
| [60]Co | 15 wt% Cr<br>18 wt% Ni<br>8 wt% Al,<br>1 wt% Mo,<br>1/2 wt% Zr | 800-1000 | S.D. | 65.5 | 53 | 367 | (1960) |
| [60]Co | 15 wt% Cr<br>18 wt% Ni<br>8 wt% Al,<br>1 wt% Nb | 800-1000 | S.D. | 61.5 | 20.4 | 367 | (1960) |

| Solute | Composition | Temperature range (°C) | Form of analysis | Activation energy, Q (kcal/mole) | Frequency factor, $D_0$ (cm²/sec) | Reference No. | Year |
|---|---|---|---|---|---|---|---|

IRON (continued)

| Solute | Composition | Temperature range (°C) | Form of analysis | Activation energy, Q (kcal/mole) | Frequency factor, $D_0$ (cm²/sec) | Reference No. | Year |
|---|---|---|---|---|---|---|---|
| $^{60}$Co | 15 wt% Cr 18 wt% Ni 8 wt% Al, 1 wt% Zr | 800-1000 | S.D. | 57.1 | 1.97 | 367 | (1960) |
| $^{60}$Co | 15 wt% Cr 18 wt% Ni 8 wt% Al, 1/2 wt% Nb, 1/2 wt% Zr | 800-1000 | S.D. | 65.5 | 92.5 | 367 | (1960) |
| $^{51}$Cr | 0.8 wt% C 0.4 wt% Mn | 1050-1250 | R.A. | 75.0 | 10 | 367 | (1954) |
| $^{51}$Cr | 12.3 wt% Cr | 900-1400 | R.A. | 55.1 | 1.29 | 369 | (1964) |
| $^{51}$Cr | 17.4 wt% Cr | 900-1400 | R.A. | 52.5 | 0.46 | 369 | (1964) |
| $^{51}$Cr | 26.0 at% Cr | 950-1275 | R.A. | 48.5 | 0.156 | 352 | (1960) |
| $^{51}$Cr | 15 wt% Cr | | R.A. | 57.0 | 2.0 | 368 | (1968) |
| $^{51}$Cr | 1.7 wt% V | | R.A. | 57.0 | 2.0 | 368 | (1968) |
| $^{64}$Cu | 0.6% Mn 0.1% Cu | 800-1200 | S.D. | 61.0 | 3.0 | 370 | (1955) |
| $^{59}$Fe | 0.27 wt% Al | 700-900 | | 52.0 | 0.17 | 371 | (1958) |
| $^{59}$Fe | 0.27 wt% Al | 1000-1250 | | 75.0 | 4.27 | 371 | (1958) |
| $^{59}$Fe | 0.39 wt% Al | 700-900 | | 53.0 | 0.29 | 371 | (1958) |
| $^{59}$Fe | 0.39 wt% Al | 1000-1250 | | 75.0 | 5.13 | 371 | (1958) |
| $^{55}$Fe | 0.0001 at% Bi | 750-890 | | 70.34 | 414.0 | 372 | (1965) |
| $^{55}$Fe | 0.0006 at% Bi | 750-890 | | 66.23 | 80.8 | 372 | (1965) |
| $^{55}$Fe | 0.0010 at% Bi | 750-890 | | 62.27 | 13.97 | 372 | (1965) |
| $^{55}$Fe | 0.0013 at% Bi | 750-890 | | 59.27 | 3.65 | 372 | (1965) |
| $^{59}$Fe | 0.1 wt% C | 720-800 | | 59.0 | 2.24 | 129 | (1960) |
| $^{59}$Fe | 0.14 wt% C | 950-1300 | R.A. | 58.0 | 0.07 | 373 | (1951) |
| $^{59}$Fe | 0.18 wt% C | 1000-1250 | R.A. | 58.0 | 0.7 | 373 | (1951) |
| $^{59}$Fe | 0.25 wt% C | 1100-1300 | S.S. | 59.0 | $5.2 \times 10^{-2}$ | 374 | (1956) |
| $^{59}$Fe | 0.54 wt% C | 1090-1300 | S.S. | 54.0 | $1.5 \times 10^{-2}$ | 374 | (1956) |
| $^{59}$Fe | 0.75 wt% C | 1090-1300 | S.S. | 54.0 | $2.1 \times 10^{-2}$ | 374 | (1956) |
| $^{59}$Fe | 1.11 wt% C | 1000-1300 | S.S. | 53.8 | $2.9 \times 10^{-2}$ | 374 | (1956) |
| $^{59}$Fe | 1.4 wt% C | 1000-1260 | S.S. | 53.8 | $5 \times 10^{-2}$ | 374 | (1956) |
| $^{59}$Fe | 2.09 wt% C | 1000-1250 | R.A. | 50.0 | $8 \times 10^{-3}$ | 373 | (1951) |
| $^{59}$Fe | 3.40 wt% C | 950-1150 | R.A. | 45.0 | $1 \times 10^{-3}$ | 373 | (1951) |
| $^{59}$Fe | 3.70 wt% C | 950-1150 | R.A. | 33.0 | $1 \times 10^{-5}$ | 373 | (1951) |
| $^{59}$Fe | 9.1 wt% Cr | 575-725 | | 55.1 | 9.27 | 340 | (1967) |

| Solute | Composition | Temperature range (°C) | Form of analysis | Activation energy, Q (kcal/mole) | Frequency factor, $D_0$ (cm²/sec) | Reference No. | Year |
|--------|-------------|------------------------|------------------|----------------------------------|-----------------------------------|---------------|------|

IRON (continued)

| Solute | Composition | Temperature range (°C) | Form of analysis | Activation energy, Q (kcal/mole) | Frequency factor, $D_0$ (cm²/sec) | Reference No. | Year |
|--------|-------------|------------------------|------------------|----------------------------------|-----------------------------------|---------------|------|
| $^{59}$Fe | 9.1 wt%Cr | 777-825 | | 52.4 | 0.42 | 340 | (1967) |
| $^{59}$Fe | 15.2 wt%Cr | 595-677 | | 54.1 | 1.25 | 340 | (1967) |
| $^{59}$Fe | 15.2 wt%Cr | 725-777 | | 51.5 | 0.27 | 340 | (1967) |
| $^{59}$Fe | 17 wt%Cr | 900-1400 | R.A. | 55.8 | 1.34 | 369 | (1964) |
| $^{59}$Fe | 19.75 wt%Cr | 575-645 | | 51.9 | 0.65 | 340 | (1967) |
| $^{59}$Fe | 19.75 wt%Cr | 700-825 | | 51.5 | 0.22 | 340 | (1967) |
| $^{59}$Fe | 21 at%Cr | 1060-1250 | S.S. | 54 | 0.32 | 354 | (1958) |
| $^{59}$Fe | 21.0 at%Cr | 1000-1200 | R.A. | 52.2 | 0.115 | 353 | (1965) |
| $^{59}$Fe | 27 at%Cr | 1040-1400 | | 50.4 | 0.195 | 351 | (1958) |
| $^{59}$Fe | 41.5 at%Cr | 950-1250 | | 65.2 | 0.11 | 354 | (1958) |
| $^{59}$Fe | 41.5 at%Cr | 1000-1200 | R.A. | 63.3 | 5.0 | 353 | (1965) |
| $^{59}$Fe | 46.7 at%Cr | 1000-1200 | R.A. | 65.9 | 15.8 | 353 | (1965) |
| $^{59}$Fe | 50.0 at%Cr | 1000-1200 | R.A. | 68.7 | 26.6 | 353 | (1965) |
| $^{59}$Fe | 17.5 wt%Cr 11.3 wt%Ni 1.3 wt%Mo | 800-1200 | S.D. | 67.1 | 0.58 | 375 | (1955) |
| $^{59}$Fe | 19.9 wt%Cr 24.7 wt%Ni | 900-1290 | S.S. | 67.9 | 1.74 | 376 | (1968) |
| $^{59}$Fe | 0.4 wt%Mn | 1150-1350 | | 83.0 | $2 \times 10^{-4}$ | 377 | (1955) |
| $^{59}$Fe | 1.15 wt%Mn | 1150-1350 | | 91.0 | $2.5 \times 10^{-3}$ | 377 | (1955) |
| $^{59}$Fe | 2.25 wt%Mn | 1150-1350 | | 94.0 | $8 \times 10^{-3}$ | 377 | (1955) |
| $^{59}$Fe | 3.18 wt%Mn | 1150-1350 | | 96.0 | $1.05 \times 10^{-2}$ | 377 | (1955) |
| $^{59}$Fe | 4.33 wt%Mn | 1150-1350 | | 98.0 | $2.4 \times 10^{-2}$ | 377 | (1955) |
| $^{59}$Fe | 5.19 wt%Mn | 1150-1350 | | 95.0 | $1.2 \times 10^{-2}$ | 377 | (1955) |
| $^{59}$Fe | 6.06 wt%Mn | 1150-1350 | | 92.5 | $4 \times 10^{-3}$ | 377 | (1955) |
| $^{59}$Fe | 8.23 wt%Mn | 1150-1350 | | 89.0 | $1 \times 10^{-3}$ | 377 | (1955) |
| $^{59}$Fe | 10 at%Ni | 950-1200 | R.A. | 67.0 | 0.5 | 378 | (1955) |
| $^{59}$Fe | 11.8 at%Ni | 1000-1200 | R.A. | 66.6 | 1.21 | 379 | (1955) |
| $^{59}$Fe | 16 at%Ni | 950-1200 | R.A. | 63.0 | 0.2 | 378 | (1955) |
| $^{59}$Fe | 20 wt%Ni 0.02 wt%C | 1050-1350 | | 75.0 | 20.0 | 380 | (1955) |
| $^{59}$Fe | 20 wt%Ni 0.18 wt%C | 1050-1350 | | 69.0 | 2.0 | 380 | (1955) |
| $^{59}$Fe | 20 wt%Ni 0.4 wt%C | 1050-1350 | | 64.0 | 0.4 | 380 | (1955) |
| $^{59}$Fe | 20 wt%Ni 0.95 wt%C | 1050-1350 | | 47.0 | $2 \times 10^{-3}$ | 380 | (1955) |
| $^{59}$Fe | 23 at%Ni | 950-1200 | R.A. | 65.0 | 0.6 | 378 | (1955) |
| $^{59}$Fe | 24.3 at%Ni | 1000-1200 | R.A. | 62.0 | 0.35 | 379 | (1955) |

| Solute | Composition | Temperature range (°C) | Form of analysis | Activation energy, Q (kcal/mole) | Frequency factor, $D_0$ (cm²/sec) | Reference No. | Year |
|---|---|---|---|---|---|---|---|

IRON (continued)

| Solute | Composition | Temperature range (°C) | Form of analysis | Activation energy, Q (kcal/mole) | Frequency factor, $D_0$ (cm²/sec) | Reference No. | Year |
|---|---|---|---|---|---|---|---|
| $^{59}$Fe | 24.1 at% Ni 0.2 at% Nb | 1000-1200 | R.A.. | 85.2 | 1260 | 379 | (1955) |
| $^{59}$Fe | 24.1 at% Ni 0.8 at% Nb | 1000-1200 | R.A. | 89.0 | 4910 | 379 | (1955) |
| $^{59}$Fe | 24.1 at% Ni 2.25 at% Nb | 1000-1200 | R.A. | 95.5 | $3.48 \times 10^4$ | 379 | (1955) |
| $^{59}$Fe | 24.4 at% Ni 0.1 at% Ti | 1000-1200 | R.A. | 67.6 | 2.38 | 379 | (1955) |
| $^{59}$Fe | 24.4 at% Ni 2.0 at% Ti | 1000-1200 | R.A. | 92.6 | 15,000 | 379 | (1955) |
| $^{59}$Fe | 24.7 at% Ni 2.3 at% V | 1000-1200 | R.A. | 55.2 | 0.035 | 379 | (1955) |
| $^{59}$Fe | 24.7 at% Ni 4.7 at% V | 1000-1200 | R.A. | 48.0 | $2.2 \times 10^{-3}$ | 379 | (1955) |
| $^{59}$Fe | 25.0 at% Ni 0.3 at% Mo | 1000-1200 | R.A. | 68 | 3.09 | 379 | (1955) |
| $^{59}$Fe | 25.0 at% Ni 1.5 at% Mo | 1000-1200 | R.A. | 77.5 | 43.3 | 379 | (1955) |
| $^{59}$Fe | 25.0 at% Ni 2.25 at% Mo | 1000-1200 | R.A. | 96.0 | $2.2 \times 10^{-3}$ | 379 | (1955) |
| $^{59}$Fe | 25 wt% Ni 0.02 wt% C | 1050-1350 | | 79 | 70 | 380 | (1955) |
| $^{59}$Fe | 25 wt% Ni 0.53 wt% C | 1050-1350 | | 67 | 1.0 | 380 | (1955) |
| $^{59}$Fe | 25 wt% Ni 0.69 wt% C | 1050-1350 | | 65 | 1.0 | 380 | (1955) |
| $^{59}$Fe | 25 wt% Ni 0.9 wt% C | 1050-1350 | | 58 | 0.1 | 380 | (1955) |
| $^{59}$Fe | 49 at% Ni | 950-1200 | R.A. | 63.0 | 0.2 | 378 | (1955) |
| $^{55}$Fe | 0.0002 at% Pb | 750-890 | | 70.43 | 435.6 | 372 | (1965) |
| $^{55}$Fe | 0.0007 at% Pb | 750-890 | | 67.56 | 145.0 | 372 | (1965) |
| $^{55}$Fe | 0.0012 at% Pb | 750-890 | | 64.28 | 34.2 | 372 | (1965) |
| $^{55}$Fe | 0.0020 at% Pb | 750-890 | | 58.35 | 2.45 | 372 | (1965) |
| $^{55}$Fe | 0.0002% Pb | | | 71.35 | 1.507 | 382 | (1966) |
| $^{55}$Fe | 0.0007% Pb | | | 69.0 | 0.652 | 382 | (1966) |
| $^{55}$Fe | 0.0012% Pb | | | 66.5 | 0.34 | 382 | (1966) |
| $^{55}$Fe | 0.002% Pb | | | 64.3 | 0.171 | 382 | (1966) |
| $^{55}$Fe | 0.0006% Sb | | | 71.5 | 1.553 | 382 | (1966) |
| $^{55}$Fe | 0.0012% Sb | | | 69.4 | 1.170 | 382 | (1966) |

| Solute | Composition | Temperature range (°C) | Form of analysis | Activation energy, Q (kcal/mole) | Frequency factor, $D_0$ (cm²/sec) | Reference No. | Year |
|---|---|---|---|---|---|---|---|

IRON (continued)

| [55]Fe | 0.004% Sb | | | 67.1 | 0.495 | 382 | (1966) |
| [55]Fe | 0.014% Sb | | | 64.5 | 0.185 | 382 | (1966) |
| [59]Fe | 3 wt% Si | 967-1416 | R.A. | 52.2 | 0.44 | 381 | (1965) |
| [55]Fe | 0.001% Sn | 1050-1350 | | 71.15 | 1.742 | 382 | (1966) |
| [55]Fe | 0.004% Sn | 1050-1350 | | 68.4 | 0.645 | 382 | (1966) |
| [55]Fe | 0.012% Sn | 1050-1350 | | 66.5 | 0.297 | 382 | (1966) |
| [55]Fe | 0.032% Sn | 1050-1350 | | 64.3 | 0.156 | 382 | (1966) |
| [59]Fe | 1.8 at% V | 700-1500 | R.A. | 56.5 | 1.39 | 383 | (1965) |
| [99]Mo | 0.7% Mo | 750-875 | R.A. | 75.0 | $1.3 \times 10^4$ | 240 | (1966) |
| [63]Ni | 5.8 at% Ni | 1152-1400 | S.D. | 73.5 | 2.11 | 86 | (1959) |
| [63]Ni | 14.88 at% Ni | 1152-1400 | S.D. | 75.6 | 5.0 | 86 | (1959) |
| [185]W | 0.006 wt% C | 1050-1250 | R.A. | 85.7 | 1900 | 248 | (1967) |
| [185]W | 0.23 wt% C | 1050-1250 | R.A. | 80.3 | 164 | 248 | (1967) |
| [185]W | 0.56 wt% C | 1050-1250 | R.A. | 77.5 | 38 | 248 | (1967) |
| [185]W | 0.71 wt% C | 1050-1250 | R.A. | 75.4 | 13 | 248 | (1967) |
| [185]W | 1.12 wt% C | 1050-1250 | R.A. | 68.4 | 0.41 | 248 | (1967) |
| [185]W | 8.5 wt% W 4.0 wt% Cr | 1015-1190 | S.S. | 74.1 | 40 | 384 | (1968) |
| [185]W | 9.5 wt% W 4.1 wt% Cr 9.7 wt% Co | 1015-1190 | S.S. | 63.9 | 0.6 | 384 | (1968) |

LEAD

| [210]Pb | 5 at% Tl | 207-323 | S.S. | 25.75 | 1.108 | 74 | (1961) |
| [210]Pb | 10 at% Tl | 207-323 | S.S. | 25.45 | 0.88 | 74 | (1961) |
| [210]Pb | 20 at% Tl | 207-323 | S.S. | 25.05 | 0.647 | 74 | (1961) |
| [210]Pb | 34 at% Tl | 207-323 | S.S. | 24.53 | 0.367 | 74 | (1961) |
| [210]Pb | 50 at% Tl | 207-323 | S.S. | 24.44 | 0.231 | 74 | (1961) |
| [210]Pb | 60 at% Tl | 207-231 | S.S. | 25.29 | 0.287 | 74 | (1961) |
| [210]Pb | 62 at% Tl | 207-231 | S.S. | 25.64 | 0.393 | 74 | (1961) |
| [210]Pb | 74 at% Tl | 207-231 | S.S. | 26.82 | 0.691 | 74 | (1961) |
| [210]Pb | 76 at% Tl | 207-231 | S.S. | 27.13 | 0.862 | 74 | (1961) |
| [210]Pb | 82 at% Tl | 207-231 | S.S. | 28.24 | 2.575 | 74 | (1961) |
| [210]Pb | 87 at% Tl | 207-231 | S.S. | 29.71 | 17 | 74 | (1961) |
| [204]Tl | 5 at% Tl | 207-325 | S.S. | 23.89 | 0.364 | 74 | (1961) |
| [204]Tl | 10 at% Tl | 207-325 | S.S. | 23.83 | 0.361 | 74 | (1961) |
| [204]Tl | 20 at% Tl | 207-325 | S.S. | 23.78 | 0.353 | 74 | (1961) |
| [204]Tl | 34 at% Tl | 207-325 | S.S. | 23.12 | 0.193 | 74 | (1961) |

| Solute | Composition | Temperature range (°C) | Form of analysis | Activation energy, Q (kcal/mole) | Frequency factor, $D_0$ (cm²/sec) | Reference No. | Year |
|---|---|---|---|---|---|---|---|
| **LEAD** (continued) | | | | | | | |
| $^{204}$Tl | 50 at% Tl | 207-325 | S.S. | 22.52 | 0.091 | 74 | (1961) |
| $^{204}$Tl | 60 at% Tl | 207-325 | S.S. | 23.20 | 0.126 | 74 | (1961) |
| $^{204}$Tl | 62 at% Tl | 206-321 | S.S. | 22.93 | 0.101 | 74 | (1961) |
| $^{204}$Tl | 74 at% Tl | 206-321 | S.S. | 23.86 | 0.194 | 74 | (1961) |
| $^{204}$Tl | 76 at% Tl | 206-321 | S.S. | 24.48 | 0.33 | 74 | (1961) |
| $^{204}$Tl | 82 at% Tl | 206-321 | S.S. | 25.37 | 0.957 | 74 | (1961) |
| $^{204}$Tl | 87 at% Tl | 206-321 | S.S. | 25.53 | 1.2 | 74 | (1961) |
| **MAGNESIUM** | | | | | | | |
| $^{110}$Ag | 42.8 at% Ag | 500-600 | S.S. | 28.7 | 0.051 | 385 | (1964) |
| $^{110}$Ag | 47.2 at% Ag | 500-700 | S.S. | 36.7 | 0.33 | 385 | (1964) |
| **MOLYBDENUM** | | | | | | | |
| $^{51}$Cr | 9.9 wt% Cr | 1200-1350 | | 72.7 | 4.3 | 386 | (1961) |
| $^{99}$Mo | 15 at% W | 1700-2300 | R.A. | 106 | 265 | 387 | (1963) |
| $^{99}$Mo | 25 at% W | 1700-2300 | R.A. | 95 | 62 | 388 | (1963) |
| $^{99}$Mo | 35 at% W | 1500-2400 | R.A. | 85 | 6.9 | 389 | (1965) |
| $^{185}$W | 15 at% W | 1500-2200 | R.A. | 73.0 | 1.4 | 387 | (1963) |
| $^{185}$W | 25 at% W | 1500-2200 | R.A. | 77.0 | 2.4 | 388 | (1963) |
| $^{185}$W | 35 at% W | 1500-2400 | R.A. | 92 | 28 | 389 | (1965) |
| **NICKEL** | | | | | | | |
| $^{14}$C | 0.74 wt% Cr | 500-800 | R.A. | 34 | 0.15 | 271 | (1957) |
| $^{14}$C | 4.65 wt% Cr | 500-800 | R.A. | 37 | 0.5 | 271 | (1957) |
| $^{14}$C | 5.25 wt% Co | 500-800 | R.A. | 37 | 0.4 | 271 | (1957) |
| $^{14}$C | 2.94 wt% Mo | 500-800 | R.A. | 38 | 1.0 | 271 | (1957) |
| $^{60}$Co | 47.3 at% Al | 1050-1350 | R.A. | 56.6 | $4.7 \times 10^{-2}$ | 350 | (1954) |
| $^{60}$Co | 48.5 at% Al | 1050-1350 | R.A. | 59.9 | $9.3 \times 10^{-2}$ | 350 | (1954) |
| $^{60}$Co | 49.4 at% Al | 1050-1350 | R.A. | 52.5 | $4.4 \times 10^{-3}$ | 350 | (1954) |
| $^{60}$Co | 4.3 at% Co | 1213-1368 | R.A. | 62.4 | 0.49 | 25 | (1965) |
| $^{60}$Co | 10.8 at% Co | 1163-1393 | R.A. | 63.0 | 0.66 | 25 | (1965) |
| $^{60}$Co | 21.1 at% Co | 1163-1393 | R.A. | 61.1 | 0.33 | 25 | (1965) |
| $^{60}$Co | 43.3 at% Co | 1163-1393 | R.A. | 62.6 | 0.52 | 25 | (1965) |
| $^{60}$Co | 48.6 at% Co | 701-819 | R.A. | 61.5 | 0.096 | 24 | (1962) |
| $^{60}$Co | 48.6 at% Co | 899-1192 | R.A. | 60.2 | 0.12 | 24 | (1962) |

| Solute | Composition | Temper-ature range (°C) | Form of analy-sis | Activation energy, Q (kcal/mole) | Frequency factor, $D_0$ (cm²/sec) | Reference No. | Year |
|--------|-------------|--------------------------|--------------------|-----------------------------------|-----------------------------------|---------------|------|

NICKEL (continued)

| Solute | Composition | Temper-ature range (°C) | Form of analy-sis | Activation energy, Q | Frequency factor, $D_0$ | Reference No. | Year |
|--------|-------------|--------------------------|--------------------|-----------------------|--------------------------|---------------|------|
| $^{60}$Co | 49.4 at% Co | 1110-1319 | R.A. | 59.8 | 0.18 | 25 | (1965) |
| $^{60}$Co | 3.14 at% Mo | 1050-1250 | S.D. | 66.0 | 0.078 | 390 | (1960) |
| $^{60}$Co | 7.91 at% Mo | 1050-1250 | S.D. | 69.0 | 0.21 | 390 | (1960) |
| $^{60}$Co | 10.29 at% Mo | 1050-1250 | S.D. | 79.6 | 6.35 | 390 | (1960) |
| $^{60}$Co | 12.99 at% Mo | 1050-1250 | S.D. | 66.5 | 0.086 | 390 | (1960) |
| $^{60}$Co | 16.79 at% Mo | 1050-1250 | S.D. | 65 | 0.046 | 390 | (1960) |
| $^{60}$Co | 20.56 at% Mo | 1050-1250 | S.D. | 58.5 | $4.4 \times 10^{-3}$ | 390 | (1960) |
| $^{60}$Co | 19.5 wt% Co 20.3 wt% Mn | 1000-1250 | R.A. | 60.5 | 0.86 | 356 | (1956) |
| $^{60}$Co | 0.13 wt% Si 0.03 wt% S | 700-1000 | R.A. | 48 | $4 \times 10^{-3}$ | 391 | (1955) |
| $^{51}$Cr | 6.35 at% Cr | 950-1200 | R.A. | 50.5 | 0.01 | 90 | (1963) |
| $^{51}$Cr | 11.69 at% Cr | 950-1200 | R.A. | 54.7 | 0.037 | 90 | (1963) |
| $^{51}$Cr | 19.8 wt% Cr 0.66 wt% Al 46 wt% Fe | 700-1000 | R.A. | 58 | 0.1 | 90 | (1955) |
| $^{51}$Cr | 20 at% Cr | | | 58.0 | | 392 | (1955) |
| $^{51}$Cr | 20.3 wt% Cr 2.6 wt% Ti 1.0 wt% Si, 0.82 wt% Al | 900-1000 | R.A. | 66 | 2 | 391 | (1955) |
| $^{51}$Cr | 19.7 wt% Cr 1.85 wt% Ti 1.3 wt% Fe, 0.79 wt% Si | 700-1000 | R.A. | 81 | 400 | 391 | (1955) |
| $^{51}$Cr | 19.8 wt% Cr 2.6 wt% Ti | 900-1300 | | 59.5 | 0.21 | 386 | (1961) |
| $^{51}$Cr | 20.4 wt% Cr 2.5 wt% Ti 1.0 wt% Al | 900-1200 | | 64 | 0.5 | 393 | (1957) |
| $^{51}$Cr | 10 at% Cr 0.19 at% Mn | 1100-1260 | S.S. | 66.5 | 1.4 | 32 | (1964) |
| $^{51}$Cr | 19.9 at% Cr 0.16 at% Mn | 1100-1270 | S.S. | 67.7 | 1.9 | 32 | (1964) |
| $^{59}$Fe | 20 wt% Cr 6 wt% W | 950-1250 | S.S. | 80.0 | 160 | 392 | (1960) |
| $^{64}$Cu | 13 at% Cu | 1054-1360 | S.S. | 63.0 | 1.5 | 32 | (1964) |
| $^{64}$Cu | 45.4 at% Cu | 985-1210 | S.S. | 60.3 | 2.3 | 32 | (1964) |
| $^{59}$Fe | 6 wt% Ti | 950-1250 | S.S. | 68.6 | 0.039 | 392 | (1960) |

| Solute | Composition | Temperature range (°C) | Form of analysis | Activation energy, Q (kcal/mole) | Frequency factor, $D_0$ (cm²/sec) | Reference No. | Year |
|---|---|---|---|---|---|---|---|
| NICKEL (continued) | | | | | | | |
| [59]Fe | 8 wt% Ti | 950-1250 | S.S. | 73.1 | 16 | 392 | (1960) |
| [59]Fe | 10.6 wt% Ti | 950-1250 | S.S. | 71.2 | 6.8 | 392 | (1960) |
| [59]Fe | 14 wt% Ti | 950-1250 | S.S. | 73 | 15 | 392 | (1960) |
| [59]Fe | 20 wt% Cr 6 wt% W 4.5 wt% Al | 950-1250 | S.S. | 71.8 | 3.1 | 392 | (1960) |
| [59]Fe | 20 wt% Cr 6 wt% W 4.5 wt% Al, 0.5 wt% Ti | 950-1250 | S.S. | 69.7 | | 392 | (1960) |
| [59]Fe | 20 wt% Cr 6 wt% W 4.5 wt% Al, 1.0 wt% Ti | 950-1250 | S.S. | 70.4 | 8.8 | 392 | (1960) |
| [59]Fe | 20 wt% Cr 6 wt% W 4.5 wt% Al, 2.0 wt% Ti | 950-1250 | S.S. | 76.1 | | 392 | (1960) |
| [59]Fe | 20 wt% Cr 6 wt% W 4.5 wt% Al, 3.0 wt% Ti | 950-1250 | S.S. | 87.6 | 1700 | 392 | (1960) |
| [59]Fe | 20 wt% Cr 6 wt% W 4.5 wt% Al, 5.0 wt% Ti | 950-1250 | S.S. | 85.5 | 1000 | 392 | (1960) |
| [59]Fe | 20 wt% Cr 6 wt% W 4.5 wt% Al, 7.0 wt% Ti | 950-1250 | S.S. | 80.0 | 160 | 392 | (1960) |
| [59]Fe | 20 wt% Cr 6 wt% W 4.5 wt% Al, 9.0 wt% Ti | 950-1250 | S.S. | 76.3 | 57 | 392 | (1960) |
| [59]Fe | 19 at% Fe | 950-1250 | R.A. | 53.0 | $4 \times 10^{-3}$ | 378 | (1955) |
| [59]Fe | 26 at% Fe | 950-1250 | R.A. | 56.0 | 0.04 | 378 | (1955) |
| [59]Fe | 33 at% Fe | 950-1200 | R.A. | 57.0 | 0.03 | 378 | (1955) |
| [59]Fe | 4 wt% Ti | 950-1250 | S.S. | 62.8 | 0.0155 | 392 | (1960) |
| [54]Mn | 13.81 wt% Mn, 2.01 wt% Sn | | | 54.4 | 0.55 | 394 | (1954) |

| Solute | Composition | Temperature range (°C) | Form of analysis | Activation energy, Q (kcal/mole) | Frequency factor, $D_0$ (cm²/sec) | Reference No. | Year |
|--------|-------------|------------------------|------------------|----------------------------------|-----------------------------------|---------------|------|

NICKEL (continued)

| Solute | Composition | Temp. range (°C) | Form | Q | $D_0$ | No. | Year |
|--------|-------------|------------------|------|---|-------|-----|------|
| $^{54}$Mn | 13.86 wt% Mn 3.67 wt% Sn | | | 70.85 | 130 | 394 | (1954) |
| $^{54}$Mn | 15.88 wt% Mn 1.83 wt% Cu | | | 62.7 | 0.21 | 394 | (1954) |
| $^{54}$Mn | 16.06 wt% Mn 0.8 wt% Ti | | | 64.9 | 1.09 | 394 | (1954) |
| $^{54}$Mn | 16.57 wt% Mn 2.13 wt% Al | | | 66.9 | 1.39 | 394 | (1954) |
| $^{54}$Mn | 16.82 wt% Mn 5.15 wt% Sn | | | 65.8 | 37.4 | 394 | (1954) |
| $^{54}$Mn | 16.93 wt% Mn 1.7 wt% Mo | | | 72.7 | 9.23 | 394 | (1954) |
| $^{63}$Ni | 10 at% Au | 999-1057 | S.S. | 51.0 | 0.04 | 364 | (1957) |
| $^{63}$Ni | 20 at% Au | 940-999 | S.S. | 49.2 | 0.05 | 364 | (1957) |
| $^{63}$Ni | 35 at% Au | 882-940 | S.S. | 39.6 | 0.005 | 364 | (1957) |
| $^{63}$Ni | 50 at% Au | 876-925 | S.S. | 44.5 | 0.09 | 364 | (1957) |
| $^{63}$Ni | 4.3 at% Co | 1213-1368 | R.A. | 68.0 | 2.86 | 25 | (1965) |
| $^{63}$Ni | 10.8 at% Co | 1166-1410 | R.A. | 60.0 | 1.47 | 25 | (1965) |
| $^{63}$Ni | 21.1 at% Co | 1166-1410 | R.A. | 62.6 | 0.45 | 25 | (1965) |
| $^{63}$Ni | 43.3 at% Co | 1163-1410 | R.A. | 63.8 | 0.69 | 25 | (1965) |
| $^{63}$Ni | 48.6 at% Co | 701-819 | R.A. | 63.6 | 0.36 | 24 | (1962) |
| $^{63}$Ni | 48.6 at% Co | 899-1192 | R.A. | 60.6 | 0.21 | 24 | (1962) |
| $^{63}$Ni | 49.4 at% Co | 1104-1347 | R.A. | 61.1 | 0.25 | 25 | (1965) |
| $^{63}$Ni | 10 at% Cr | 1040-1275 | S.S. | 70.2 | 3.3 | 32 | (1964) |
| $^{63}$Ni | 19.9 at% Cr | 1040-1275 | S.S. | 68.2 | 1.6 | 32 | (1964) |
| $^{63}$Ni | 29.7 at% Cr | 1040-1275 | S.S. | 70.5 | 2.9 | 32 | (1964) |
| $^{63}$Ni | 13 at% Cu | 1054-1360 | S.S. | 74.9 | 35 | 32 | (1964) |
| $^{63}$Ni | 45.4 at% Cu | 985-1210 | S.S. | 66.8 | 17 | 32 | (1964) |
| $^{63}$Ni | 20% Fe | 814-1200 | | 52.3 | $7.19 \times 10^{-3}$ | 395 | (1967) |
| $^{63}$Ni | 26% Fe | 814-1200 | | 49.7 | $1.76 \times 10^{-3}$ | 395 | (1967) |
| $^{63}$Ni | 27% Fe | 814-1200 | | 48.7 | $1.95 \times 10^{-3}$ | 395 | (1967) |
| $^{63}$Ni | 30% Fe | 814-1200 | | 51.9 | $6.8 \times 10^{-3}$ | 395 | (1967) |
| $^{63}$Ni | 35% Fe | 814-1200 | | 54.8 | $1.67 \times 10^{-2}$ | 395 | (1967) |
| $^{63}$Ni | 1.2 wt% Nb | 1030-1230 | S.S. | 60.8 | 0.12 | 396 | (1968) |
| $^{63}$Ni | 8.0 wt% Nb | 1030-1230 | S.S. | 62.2 | 0.20 | 396 | (1968) |
| $^{63}$Ni | 10.0 wt% Nb | 1030-1230 | S.S. | 67.1 | 1.0 | 396 | (1968) |
| $^{63}$Ni | 1.7 at% W | 1100-1300 | S.S. | 76.5 | 30 | 32 | (1964) |
| $^{63}$Ni | 5.3 at% W | 1100-1300 | S.S. | 80.6 | 58 | 32 | (1964) |
| $^{63}$Ni | 9.2 at% W | 1100-1300 | S.S. | 70.3 | 1.1 | 32 | (1964) |

| Solute | Composition | Temperature range (°C) | Form of analysis | Activation energy, Q (kcal/mole) | Frequency factor, $D_0$ (cm²/sec) | Reference No. | Year |
|---|---|---|---|---|---|---|---|

**NICKEL** (continued)

| Solute | Composition | Temperature range (°C) | Form of analysis | Activation energy, Q | Frequency factor, $D_0$ | Ref. No. | Year |
|---|---|---|---|---|---|---|---|
| $^{185}$W | 1.7 at% W | 1100-1300 | S.S. | 73.1 | 2.2 | 32 | (1964) |
| $^{185}$W | 5.3 at% W | 1100-1300 | S.S. | 80.5 | 17 | 32 | (1964) |
| $^{185}$W | 9.2 at% W | 1100-1300 | S.S. | 74.5 | 1.4 | 32 | (1964) |

**NIOBIUM**

| $^{95}$Nb | 10.0% Mo | 1600-2100 | R.A. | 126.0 | 26.9 | 397 | (1966) |
| $^{95}$Nb | 20.0% Mo | 1700-2100 | R.A. | 112.0 | 60.3 | 397 | (1966) |
| $^{95}$Nb | 30.0% Mo | 1700-2100 | R.A. | 105.0 | 5.26 | 397 | (1966) |
| $^{95}$Nb | 45.0% Mo | 1700-2100 | R.A. | 94.0 | 0.27 | 397 | (1966) |
| $^{95}$Nb | 11 at% Ti | 1720-2160 | A.R.G. | 91.0 | 1.0 | 398 | (1962) |
| $^{95}$Nb | 34 at% Ti | 1490-1960 | A.R.G. | 73.0 | 0.15 | 398 | (1962) |
| $^{95}$Nb | 47 at% Ti | 1240-1790 | A.R.G. | 64.8 | 0.08 | 398 | (1962) |
| $^{95}$Nb | 5% W | 1600-2000 | | 130 | 2334 | 285 | (1967) |
| $^{95}$Nb | 10% W | 1600-2000 | | 117 | 164 | 285 | (1967) |
| $^{95}$Nb | 30% W | 1600-2000 | | 85 | $2.57 \times 10^{-2}$ | 285 | (1967) |

**PLATINUM**

| $^{64}$Cu | 25.5 at% Cu | 1098-1385 | S.S. | 64.4 | 0.67 | 214 | (1963) |
| $^{64}$Pt | 25.5 at% Cu | 1140-1382 | S.S. | 60.3 | 0.022 | 214 | (1963) |

**SILVER**

| $^{110}$Ag | 2.05 at% Al | 700-850 | S.S. | 42.5 | 0.25 | 306 | (1955) |
| $^{110}$Ag | 9.47 at% Al | 700-850 | S.S. | 42.9 | 0.83 | 306 | (1955) |
| $^{110}$Ag | 14.1 at% Al | 700-850 | S.S. | 41.2 | 0.73 | 306 | (1955) |
| $^{110}$Ag | 8 at% Au | 654-945 | S.S. | 44.79 | 0.52 | 363 | (1963) |
| $^{110}$Ag | 17 at% Au | 634-952 | S.S. | 44.05 | 0.32 | 363 | (1963) |
| $^{110}$Ag | 35 at% Au | 635-956 | S.S. | 43.54 | 0.23 | 363 | (1963) |
| $^{110}$Ag | 38.1 wt% Au | 656-910 | S.S. | 42.8 | 0.064 | 42 | (1957) |
| $^{110}$Ag | 50 at% Au | 634-972 | S.S. | 43.11 | 0.19 | 363 | (1963) |
| $^{110}$Ag | 0.9 wt% Cd | 500-800 | R.A. | 41.5 | 0.18 | 399 | (1957) |
| $^{110}$Ag | 1.09 wt% Cd | 500-800 | R.A. | 39.9 | 0.13 | 399 | (1957) |
| $^{110}$Ag | 2.8 wt% Cd | 500-800 | R.A. | 38.3 | 0.06 | 399 | (1957) |
| $^{110}$Ag | 5.13 wt% Cd | 500-800 | R.A. | 37.0 | 0.03 | 399 | (1957) |
| $^{110}$Ag | 6.5 at% Cd | 571-908 | S.S. | 42.61 | 0.306 | 400 | (1958) |

| Solute | Composition | Temperature range (°C) | Form of analysis | Activation energy, Q (kcal/mole) | Frequency factor, $D_0$ (cm²/sec) | Reference No. | Year |
|---|---|---|---|---|---|---|---|

SILVER (continued)

| Solute | Composition | Temperature range (°C) | Form of analysis | Activation energy, Q (kcal/mole) | Frequency factor, $D_0$ (cm²/sec) | Reference No. | Year |
|---|---|---|---|---|---|---|---|
| $^{110}$Ag | 13.6 at% Cd | 558-850 | S.S. | 40.96 | 0.234 | 400 | (1958) |
| $^{110}$Ag | 22 wt% Cd | | | 35.0 | 0.03 | 399 | (1957) |
| $^{110}$Ag | 28 at% Cd | 505-740 | S.S. | 37.25 | 0.156 | 400 | (1958) |
| $^{110}$Ag | 0.17 wt% Cu | 780-906 | R.A. | 45.2 | 0.65 | 126 | (1957) |
| $^{110}$Ag | 0.17 wt% Cu | 690-780 | R.A. | 41.0 | 1.06 | 126 | (1957) |
| $^{110}$Ag | 0.84 wt% Cu | 780-906 | R.A. | 45.2 | 0.68 | 126 | (1957) |
| $^{110}$Ag | 0.84 wt% Cu | 690-780 | R.A. | 39.0 | 0.08 | 126 | (1957) |
| $^{110}$Ag | 1.68 wt% Cu | 690-780 | R.A. | 40.6 | 0.07 | 126 | (1957) |
| $^{110}$Ag | 1.75 at% Cu | 750-850 | S.S. | 44.8 | 0.66 | 306 | (1955) |
| $^{110}$Ag | 4.16 at% Cu | 750-890 | S.S. | 46.6 | 1.84 | 306 | (1955) |
| $^{110}$Ag | 5.0 wt% Cu | 700-840 | R.A. | 39.9 | 0.06 | 126 | (1957) |
| $^{110}$Ag | 6.56 at% Cu | 700-840 | S.S. | 43.5 | 0.51 | 306 | (1955) |
| $^{110}$Ag | 8.15 wt% Cu | 725-830 | R.A. | 38.3 | 0.04 | 126 | (1957) |
| $^{110}$Ag | 1.50 at% Ge | 700-850 | S.S. | 44.0 | 0.55 | 306 | (1955) |
| $^{110}$Ag | 3.0 at% Ge | 700-850 | S.S. | 45.3 | 1.59 | 306 | (1955) |
| $^{110}$Ag | 4.3 at% Ge | 700-800 | S.S. | 44.5 | 1.89 | 306 | (1955) |
| $^{110}$Ag | 5.43 at% Ge | 700-770 | S.S. | 44.2 | 2.18 | 306 | (1955) |
| $^{110}$Ag | 4.4 at% In | 646-888 | S.S. | 42.67 | 0.358 | 400 | (1958) |
| $^{110}$Ag | 12.6 at% In | 573-795 | S.S. | 37.40 | 0.116 | 400 | (1958) |
| $^{110}$Ag | 16.7 at% In | 576-728 | S.S. | 36.27 | 0.183 | 400 | (1958) |
| $^{110}$Ag | 41.1 at% Mg | 500-700 | S.S. | 33.2 | 0.095 | 385 | (1964) |
| $^{110}$Ag | 43.6 at% Mg | 500-700 | S.S. | 35.3 | 0.15 | 385 | (1964) |
| $^{110}$Ag | 48.5 at% Mg | 500-700 | S.S. | 39.5 | 0.37 | 385 | (1964) |
| $^{110}$Ag | 48.7 at% Mg | 500-700 | S.S. | 39.7 | 0.39 | 385 | (1964) |
| $^{110}$Ag | 0.21 at% Pb | 600-800 | S.S. | 42.5 | 0.22 | 306 | (1955) |
| $^{110}$Ag | 0.71 at% Pb | 650-850 | S.S. | 44.7 | 0.89 | 306 | (1955) |
| $^{110}$Ag | 1.30 at% Pb | 650-850 | S.S. | 43.5 | 0.70 | 306 | (1955) |
| $^{110}$Ag | 1.49 at% Pd | 715-942 | S.S. | 43.7 | 0.239 | 401 | (1957) |
| $^{110}$Ag | 3.69 at% Pd | 715-942 | S.S. | 43.7 | 0.194 | 401 | (1957) |
| $^{110}$Ag | 9.87 at% Pd | 715-942 | S.S. | 43.7 | 0.122 | 401 | (1957) |
| $^{110}$Ag | 21.8 at% Pd | 715-942 | S.S. | 43.7 | 0.043 | 401 | (1957) |
| $^{110}$Ag | 0.5 wt% Sb | | | 39.6 | 0.304 | 402 | (1956) |
| $^{110}$Ag | 0.53 at% Sb | 700-900 | S.S. | 43.5 | 0.382 | 403 | (1955) |
| $^{110}$Ag | 0.6 wt% Sb | | | 39.6 | 0.1 | 399 | (1957) |
| $^{110}$Ag | 0.89 at% Sb | 570-890 | S.S. | 42.6 | 0.302 | 403 | (1955) |
| $^{110}$Ag | 1.42 at% Sb | 568-891 | S.S. | 42.0 | 0.275 | 403 | (1955) |
| $^{110}$Ag | 3.65 wt% Sb | | | 27.8 | $1.2 \times 10^{-3}$ | 399 | (1957) |
| $^{110}$Ag | 5.2 wt% Sb | | | 29.8 | $5.6 \times 10^{-3}$ | 399 | (1957) |
| $^{110}$Ag | 5.2 wt% Sb | | | 28.8 | $7.4 \times 10^{-3}$ | 399 | (1957) |

| Solute | Composition | Temper-ature range (°C) | Form of analy-sis | Activation energy, Q (kcal/mole) | Frequency factor, $D_0$ (cm²/sec) | Reference No. | Year |
|---|---|---|---|---|---|---|---|
| **SILVER** (continued) | | | | | | | |
| $^{110}$Ag | 0.18 wt% Sn | 830-900 | R.A. | 45.2 | 0.622 | 126 | (1957) |
| $^{110}$Ag | 0.18 wt% Sn | 700-830 | R.A. | 41.7 | 0.132 | 126 | (1957) |
| $^{110}$Ag | 0.48 wt% Sn | 700-850 | R.A. | 40.9 | 0.128 | 126 | (1957) |
| $^{110}$Ag | 0.91 wt% Sn | 700-850 | R.A. | 40.5 | 0.17 | 126 | (1957) |
| $^{110}$Ag | 0.97 wt% Sn | | | 39.4 | 0.28 | 399 | (1957) |
| $^{110}$Ag | 1.0 wt% Sn | | | 39.8 | 0.47 | 402 | (1956) |
| $^{110}$Ag | 2.8 wt% Sn | 700-850 | R.A. | 40.3 | 0.20 | 399 | (1957) |
| $^{110}$Ag | 4.6 wt% Sn | 700-830 | R.A. | 39.7 | 0.225 | 399 | (1957) |
| $^{110}$Ag | 5.1 wt% Sn | | | 38.6 | 0.2 | 399 | (1957) |
| $^{110}$Ag | 7.4 wt% Sn | | | 37.0 | 0.16 | 399 | (1957) |
| $^{110}$Ag | 7.5 wt% Sn | | | 37.3 | 0.56 | 402 | (1956) |
| $^{110}$Ag | 1.1 at% Tl | 640-870 | S.S. | 43.5 | 0.42 | 302 | (1958) |
| $^{110}$Ag | 2.6 at% Tl | 640-870 | S.S. | 44.9 | 0.35 | 302 | (1958) |
| $^{110}$Ag | 5.5 at% Tl | 640-870 | S.S. | 37.6 | 0.10 | 302 | (1958) |
| $^{110}$Ag | 5 wt% Zn | 640-925 | S.S. | 41.7 | 0.54 | 404 | (1950) |
| $^{110}$Ag | 30 at% Zn | 773-971 | S.S. | 35.99 | 0.29 | 405 | (1956) |
| $^{115}$Cd | 6.5 at% Cd | 571-922 | S.S. | 40.48 | 0.328 | 400 | (1958) |
| $^{115}$Cd | 13.6 at% Cd | 522-867 | S.S. | 38.61 | 0.218 | 400 | (1958) |
| $^{115}$Cd | 27.5 at% Cd | 526-795 | S.S. | 35.95 | 0.253 | 400 | (1958) |
| $^{114}$In | 4.7 at% In | 573-795 | S.S. | 40.30 | 0.453 | 400 | (1958) |
| $^{114}$In | 12.4 at% In | 573-888 | S.S. | 38.39 | 0.566 | 400 | (1958) |
| $^{114}$In | 16.6 at% In | 474-729 | S.S. | 36.61 | 0.537 | 400 | (1958) |
| $^{210}$Pb | 0.21 at% Pb | 700-810 | S.S. | 42.5 | 0.22 | 406 | (1952) |
| $^{210}$Pb | 0.25 at% Pb | 700-850 | S.S. | 37.8 | 0.22 | 306 | (1955) |
| $^{210}$Pb | 0.52 at% Pb | 700-850 | S.S. | 38.7 | 0.38 | 306 | (1955) |
| $^{210}$Pb | 0.71 at% Pb | 700-810 | S.S. | 44.7 | 0.89 | 406 | (1952) |
| $^{210}$Pb | 1.30 at% Pb | 700-810 | S.S. | 43.5 | 0.70 | 406 | (1952) |
| $^{210}$Pb | 1.32 at% Pb | 700-850 | S.S. | 38.5 | 0.46 | 306 | (1955) |
| $^{124}$Sb | 0.89 at% Sb | 570-710 | S.S. | 38.3 | 0.17 | 403 | (1955) |
| $^{204}$Tl | 1.1 at% Tl | 640-870 | S.S. | 40.4 | 0.72 | 302 | (1958) |
| $^{204}$Tl | 2.6 at% Tl | 640-870 | S.S. | 39.4 | 0.57 | 302 | (1958) |
| $^{65}$Zn | 30 at% Zn | 768-969 | S.S. | 35.21 | 0.46 | 405 | (1958) |
| **TITANIUM** | | | | | | | |
| $^{51}$Cr | 10 at% Cr | 900-1200 | A.R.G. | 40.2 | 0.02 | 324 | (1959) |
| $^{51}$Cr | 18 at% Cr | 900-1200 | A.R.G. | 44.5 | 0.09 | 324 | (1959) |
| $^{55}$Fe | 5 at% Fe | 850-1250 | A.R.G. | 39.6 | 0.092 | 326 | (1962) |
| $^{55}$Fe | 10 at% Fe | 850-1250 | A.R.G. | 48.6 | 2.14 | 326 | (1962) |

| Solute | Composition | Temperature range (°C) | Form of analysis | Activation energy, Q (kcal/mole) | Frequency factor, $D_0$ (cm²/sec) | Reference No. | Year |
|---|---|---|---|---|---|---|---|
| **TITANIUM** (continued) | | | | | | | |
| $^{55}$Fe | 15 at% Fe | 850-1100 | A.R.G. | 58.1 | 52.5 | 326 | (1962) |
| $^{55}$Fe | 5 at% Nb | 850-1250 | A.R.G. | 33.1 | $7.9 \times 10^{-3}$ | 326 | (1962) |
| $^{55}$Fe | 10 at% Nb | 850-1300 | A.R.G. | 34.9 | $1.15 \times 10^{-3}$ | 326 | (1962) |
| $^{55}$Fe | 15 at% Nb | 850-1250 | A.R.G. | 35.0 | $7.9 \times 10^{-3}$ | 326 | (1962) |
| $^{95}$Nb | 5 at% Fe | 1000-1250 | A.R.G. | 34.9 | $1.82 \times 10^{-4}$ | 326 | (1962) |
| $^{95}$Nb | 10 at% Fe | 900-1250 | A.R.G. | 41.3 | $2.9 \times 10^{-2}$ | 326 | (1962) |
| $^{95}$Nb | 15 at% Fe | 900-1100 | A.R.G. | 56.0 | 9.9 | 326 | (1962) |
| $^{95}$Nb | 5 at% Nb | 950-1250 | A.R.G. | 29.9 | $1.2 \times 10^{-4}$ | 326 | (1962) |
| $^{95}$Nb | 10 at% Nb | 800-1250 | A.R.G. | 36.1 | $6.8 \times 10^{-4}$ | 326 | (1962) |
| $^{95}$Nb | 15 at% Nb | 880-1260 | A.R.G. | 39.3 | $1.5 \times 10^{-3}$ | 326 | (1962) |
| $^{44}$Ti | 10 wt% V | 900-1600 | S.S. | Nonlinear | | 331 | (1968) |
| $^{44}$Ti | 20 wt% V | 900-1550 | S.S. | Nonlinear | | 331 | (1968) |
| $^{44}$Ti | 30 wt% V | 950-1575 | S.S. | Nonlinear | | 331 | (1968) |
| $^{44}$Ti | 40 wt% V | 1000-1575 | S.S. | Nonlinear | | 331 | (1968) |
| $^{44}$Ti | 50 wt% V | 1000-1575 | S.S. | Nonlinear | | 331 | (1968) |
| $^{48}$V | 10 wt% V | 900-1575 | S.S. | Nonlinear | | 331 | (1968) |
| $^{48}$V | 20 wt% V | 900-1575 | S.S. | Nonlinear | | 331 | (1968) |
| $^{48}$V | 30 wt% V | 950-1575 | S.S. | Nonlinear | | 331 | (1968) |
| $^{48}$V | 40 wt% V | 950-1575 | S.S. | Nonlinear | | 331 | (1968) |
| $^{48}$V | 50 wt% V | 1050-1575 | S.S. | Nonlinear | | 331 | (1968) |
| $^{95}$Nb | 34 at% Nb | 1060-1750 | A.R.G. | 50.5 | $9.6 \times 10^{-3}$ | 398 | (1962) |
| $^{46}$Sc | 10 wt% Nb | 1000-1200 | | 15.77 | − | 407 | (1957) |
| $^{46}$Sc | 20 wt% Nb | 1000-1200 | | 30.15 | − | 407 | (1957) |
| $^{46}$Sc | 30 wt% Nb | 1000-1200 | | 31.50 | − | 407 | (1957) |
| $^{46}$Sc | 40 wt% Nb | 1000-1200 | | 13.50 | − | 407 | (1957) |
| $^{46}$Sc | 50 wt% Nb | 1000-1200 | | 4.77 | − | 407 | (1957) |
| **TUNGSTEN** | | | | | | | |
| $^{99}$Mo | 25 at% Mo | 1900-2500 | R.A. | 83 | 0.18 | 388 | (1963) |
| $^{186}$Re | 27% Re | 2000-2400 | | 94.0 | $9.2 \times 10^{-3}$ | 332 | (1967) |
| $^{185}$W | 2 wt% Be | 900-1200 | | 33.8 | $1.18 \times 10^{-3}$ | 408 | (1962) |
| $^{185}$W | 12 wt% Be | 900-1200 | | 66.95 | 2.36 | 408 | (1962) |
| $^{185}$W | 24 wt% Be | 900-1200 | | 33.12 | $1.1 \times 10^{-4}$ | 408 | (1962) |
| $^{185}$W | 25 at% Mo | 2100-2600 | R.A. | 116 | 26 | 389 | (1963) |
| **URANIUM** | | | | | | | |
| $^{235}$U | 4 wt% Mo | 800-1040 | R.A. | 33.0 | $2.5 \times 10^{-3}$ | 409 | (1961) |

| Solute | Composition | Temperature range (°C) | Form of analysis | Activation energy, Q (kcal/mole) | Frequency factor, $D_0$ (cm²/sec) | Reference No. | Year |
|--------|-------------|------------------------|------------------|----------------------------------|-----------------------------------|---------------|------|
| **URANIUM** (continued) | | | | | | | |
| $^{235}$U | 4 wt% Nb | 800-1040 | R.A. | 28.2 | $1.66 \times 10^{-4}$ | 409 | (1961) |
| $^{235}$U | 4 wt% Zr | 800-1040 | R.A. | 22.0 | $1.26 \times 10^{-4}$ | 409 | (1961) |
| **VANADIUM** | | | | | | | |
| $^{44}$Ti | 10 wt% | 1150-1750 | S.S. | | Nonlinear | 331 | (1968) |
| $^{44}$Ti | 20 wt% | 1100-1670 | S.S. | | Nonlinear | 331 | (1968) |
| $^{44}$Ti | 30 wt% | 1100-1650 | S.S. | | Nonlinear | 331 | (1968) |
| $^{44}$Ti | 40 wt% | 1000-1600 | S.S. | | Nonlinear | 331 | (1968) |
| $^{48}$V | 10 wt% | 1150-1750 | S.S. | | Nonlinear | 331 | (1968) |
| $^{48}$V | 20 wt% | 1150-1650 | S.S. | | Nonlinear | 331 | (1968) |
| $^{48}$V | 30 wt% | 1100-1600 | S.S. | | Nonlinear | 331 | (1968) |
| $^{48}$V | 40 wt% | 1050-1570 | S.S. | | Nonlinear | 331 | (1968) |
| **ZINC** | | | | | | | |
| $^{65}$Zn | 0.57 wt% Ag, ‖c, S | 345-420 | | 21.95 | 0.14 | 410 | (1967) |
| $^{65}$Zn | 0.57 wt% Ag, ⊥c, S | 345-420 | | 23.12 | 0.22 | 410 | (1967) |
| $^{65}$Zn | 1.4 wt% Ag, ‖c, S | 345-420 | | 22.07 | 0.17 | 410 | (1967) |
| $^{65}$Zn | 1.4 wt% Ag, ⊥c, S | 345-420 | | 23.13 | 0.26 | 410 | (1967) |
| $^{65}$Zn | 0.5 wt% Al, ‖c | | R.A. | 19.0 | $1.0 \times 10^{-2}$ | 411 | (1957) |
| $^{65}$Zn | 0.5 wt% Al, ⊥c | | R.A. | 26.0 | 1.56 | 411 | (1957) |
| $^{65}$Zn | 37.1 at% Al | 325-405 | S.S. | 20.0 | 0.012 | 173 | (1959) |
| $^{65}$Zn | 0.5 wt% Cu, ‖c | | R.A. | 19.5 | $3.1 \times 10^{-2}$ | 411 | (1957) |
| $^{65}$Zn | 0.5 wt% Cu, ⊥c | | R.A. | 26.0 | 3.47 | 411 | (1957) |
| $^{65}$Zn | 1 wt% Cu, ‖c | | R.A. | 19.5 | $4.0 \times 10^{-2}$ | 411 | (1957) |
| $^{65}$Zn | 1 at% Cu, ⊥c | | R.A. | 25.5 | 4.4 | 411 | (1957) |
| **ZIRCONIUM** | | | | | | | |
| $^{113}$Sn | 0.7 wt% Sn | 1000-1250 | R.A. | 43 | $2 \times 10^{-2}$ | 178 | (1959) |
| $^{113}$Sn | 3.3 wt% Sn | 1000-1250 | R.A. | 50 | 0.2 | 178 | (1959) |
| $^{113}$Sn | 5.6 wt% Sn | 1000-1250 | R.A. | 52 | 0.4 | 178 | (1959) |
| $^{95}$Zr | 1.33 wt% Nb 0.7 wt% Ta | 900-1200 | R.A. | 33 | $3 \times 10^{-4}$ | 346 | (1960) |
| $^{95}$Zr | 2 wt% Nb | 900-1200 | R.A. | 31 | $1.5 \times 10^{-4}$ | 346 | (1960) |
| $^{95}$Zr | 5 wt% Nb | 900-1200 | R.A. | 33 | $3 \times 10^{-4}$ | 346 | (1960) |
| $^{95}$Zr | 10 wt% Nb | 900-1200 | R.A. | 35 | $4 \times 10^{-4}$ | 346 | (1960) |

| Solute | Composition | Temperature range (°C) | Form of analysis | Activation energy, Q (kcal/mole) | Frequency factor, $D_0$ (cm²/sec) | Reference No. | Year |
|--------|-------------|------------------------|------------------|----------------------------------|-----------------------------------|---------------|------|

ZIRCONIUM (continued)

| Solute | Composition | Temperature range (°C) | Form of analysis | Activation energy, Q (kcal/mole) | Frequency factor, $D_0$ (cm²/sec) | Reference No. | Year |
|--------|-------------|------------------------|------------------|----------------------------------|-----------------------------------|---------------|------|
| $^{95}$Zr | 0.7 wt% Sn | 1000-1250 | R.A. | 36 | $2 \times 10^{-3}$ | 412 | (1959) |
| $^{95}$Zr | 1.3 wt% Sn | 740-827 | R.A. | 62 | 5.0 | 412 | (1959) |
| $^{95}$Zr | 2.39 wt% Sn | 740-827 | R.A. | 75 | 2100 | 412 | (1959) |
| $^{95}$Zr | 3.3 wt% Sn | 1000-1250 | R.A. | 39 | $3 \times 10^{-3}$ | 178 | (1959) |
| $^{95}$Zr | 3.54 wt% Sn | 740-827 | R.A. | 64.0 | 10 | 412 | (1959) |
| $^{95}$Zr | 5.6 wt% Sn | 1000-1250 | R.A. | 50 | 0.2 | 178 | (1959) |
| $^{95}$Zr | 2 wt% Ta | 900-1200 | R.A. | 26.5 | $3 \times 10^{-5}$ | 344 | (1960) |
| $^{95}$Zr | 5 wt% Ta | 900-1200 | R.A. | 28.0 | $6 \times 10^{-5}$ | 344 | (1960) |
| $^{95}$Zr | 10 wt% Ta | 900-1200 | R.A. | 30.5 | $1.5 \times 10^{-4}$ | 344 | (1960) |

## Part IV

## Self- and Impurity Diffusion in Simple Metal Oxides

| Tracer | Material | Temperature range (°C) | Activation energy, Q (kcal/mole) | Frequency factor, $D_0$ (cm²/sec) | Reference | |
|--------|----------|------------------------|----------------------------------|-----------------------------------|-----------|------|
| | | | | | No. | Year |

ALUMINUM OXIDE, $Al_2O_3$

| Tracer | Material | Temperature range (°C) | Activation energy, Q (kcal/mole) | Frequency factor, $D_0$ (cm²/sec) | No. | Year |
|--------|----------|------------------------|----------------------------------|-----------------------------------|-----|--------|
| $^{26}$Al | P 99.9 | 1670-1905 | 114.0 | 28 | 413 | (1962) |
| $^{241}$Am | P 99.9 | 1200-1430 | 135.0 | $4.94 \times 10^6$ | 414 | (1968) |
| $^{59}$Fe | P | 900-1200 *** | 82.0 *** | 1.13 | 415 | (1958) |
| $^{59}$Fe | P 99.5 | 900-1100 | 27.0 | $9.18 \times 10^{-8}$ | 416 | (1958) |
| $^{147}$Pr | P | 1350-1540 | 157 | $1.18 \times 10^8$ | 417 | (1965) |
| $^{239}$Pu | P 99.9 | 1200-1450 | 142.2 | $1.54 \times 10^8$ | 414 | (1968) |

BARIUM OXIDE, BaO

| Tracer | Material | Temperature range (°C) | Activation energy, Q (kcal/mole) | Frequency factor, $D_0$ (cm²/sec) | No. | Year |
|--------|----------|------------------------|----------------------------------|-----------------------------------|-----|--------|
| $^{140}$Ba | S | 1080-1250 | 253 | $10^{29}$ | 418 | (1952) |

BERYLLIUM OXIDE, BeO

| Tracer | Material | Temperature range (°C) | Activation energy, Q (kcal/mole) | Frequency factor, $D_0$ (cm²/sec) | No. | Year |
|--------|----------|------------------------|----------------------------------|-----------------------------------|-----|--------|
| $^7$Be | P 99.97 | 1725-2000 | 36.15 | $1.1 \times 10^{-6}$ | 419 | (1961) |
| $^7$Be | P 99.97 | 1550-1725 | 91.9 | 1.37 | 419 | (1961) |
| $^7$Be | P 99.6 | 1730-1930 | 66.1 | $6.14 \times 10^{-2}$ | 420 | (1964) |
| $^7$Be | P 99.6 | 1570-1730 | 111.6 | 5560 | 420 | (1964) |
| $^7$Be | S 99.99 | 1760-2000 | 64 | $1.27 \times 10^{-3}$ | 420 | (1964) |
| $^7$Be | S 99.99 | 1490-1720 | 36 | $1.23 \times 10^{-6}$ | 420 | (1964) |
| $^7$Be | P 99.97 | 1500-1725 | 92.0 | 1.35 | 420 | (1964) |
| $^7$Be | P 99.97 | 1760-2000 | 36.0 | $1.07 \times 10^{-6}$ | 420 | (1964) |
| $^7$Be | P 99.9 | 1150-1800 | 62.5 | $2.49 \times 10^{-3}$ | 421 | (1964) |
| $^7$Be | S ⊥ c, ‖c | 1500-2000 | 53.4 | $5.8 \times 10^{-5}$ | 422 | (1966) |
| $^7$Be | P | 1100-1800 | 63.0 | $3.2 \times 10^{-3}$ | 423 | (1966) |

| Tracer | Material | Temper- ature range (°C) | Activation energy, Q (kcal/mole) | Frequency factor, $D_0$ (cm²/sec) | Reference No. | Year |
|--------|----------|----------|----------|----------|------|------|

**BISMUTH OXIDE, $Bi_2O_3$**

| Tracer | Material | Temper- ature range (°C) | Activation energy, Q (kcal/mole) | Frequency factor, $D_0$ (cm²/sec) | Reference No. | Year |
|--------|----------|----------|----------|----------|------|------|
| $^{210}Bi$ | P | 600-710 | 20.7 | $4.29 \times 10^{-6}$ | 424 | (1963) |
| $^{210}Bi$ | P | 710-780 | 66.0 | 0.45 | 424 | (1963) |

**CALCIUM OXIDE, CaO**

| Tracer | Material | Temper- ature range (°C) | Activation energy, Q (kcal/mole) | Frequency factor, $D_0$ (cm²/sec) | Reference No. | Year |
|--------|----------|----------|----------|----------|------|------|
| $^{45}Ca$ | P | 900-1400 | 81.0 | 0.4 | 425 | (1952) |
| $^{45}Ca$ | S | 1000-1400 | 34.6 | $8.75 \times 10^{-8}$ | 426 | (1967) |

**CHROMIUM OXIDE, $Cr_2O_3$**

| Tracer | Material | Temper- ature range (°C) | Activation energy, Q (kcal/mole) | Frequency factor, $D_0$ (cm²/sec) | Reference No. | Year |
|--------|----------|----------|----------|----------|------|------|
| $^{51}Cr$ | P | 1000-1350 | 100 | 4000 | 427 | (1956) |
| $^{51}Cr$ | P | 900-1100 | 22 | $4.29 \times 10^{-8}$ | 415 | (1958) |
| $^{51}Cr$ | P 99.0 | 1100-1550 | 61.0 | 0.137 | 428 | (1961) |
| $^{59}Fe$ | P | 900-1100 | 44 | $4.95 \times 10^{-6}$ | 415 | (1958) |

**COBALT OXIDE, CoO**

| Tracer | Material | Temper- ature range (°C) | Activation energy, Q (kcal/mole) | Frequency factor, $D_0$ (cm²/sec) | Reference No. | Year |
|--------|----------|----------|----------|----------|------|------|
| $^{60}Co$ | P 99.99 | 800-1350 | 34.5 | $2.15 \times 10^{-3}$ | 429 | (1954) |

**COPPER OXIDE, $Cu_2O$**

| Tracer | Material | Temper- ature range (°C) | Activation energy, Q (kcal/mole) | Frequency factor, $D_0$ (cm²/sec) | Reference No. | Year |
|--------|----------|----------|----------|----------|------|------|
| $^{110}Ag$ | S | 800-1050 *** | 27.63 *** | $6 \times 10^{-3}$ | 430 | (1960) |
| $^{110}Ag$ | P | 700-800 | 14.0 | $3.8 \times 10^{-3}$ | 430 | (1960) |
| $^{110}Ag$ | P | 800-1000 | 28.27 | $5.6 \times 10^{-3}$ | 430 | (1960) |
| $^{64}Cu$ | P | 800-1050 | 36.1 | $4.36 \times 10^{-2}$ | 431 | (1951) |
| $^{114}In$ | S | 800-1050 | 33.5 | 160 | 432 | (1962) |
| $^{114}In$ | P | 600-780 | 12.4 | $2.4 \times 10^{-8}$ | 426 | (1962) |
| $^{114}In$ | P | 780-1050 | 24.8 | $8.9 \times 10^{-6}$ | 432 | (1962) |
| $^{65}Zn$ | P | 600-900 | 9.0 | $1.7 \times 10^{-8}$ | 433 | (1960) |
| $^{65}Zn$ | P | 900-1050 | 30.9 | $5.2 \times 10^{-4}$ | 433 | (1960) |

**IRON OXIDE, FeO**

| Tracer | Material | Temper- ature range (°C) | Activation energy, Q (kcal/mole) | Frequency factor, $D_0$ (cm²/sec) | Reference No. | Year |
|--------|----------|----------|----------|----------|------|------|
| $^{59}Fe$ | P | 700-1000 | 29.7 | 0.118 | 434 | (1953) |
| $^{59}Fe$ | P | 700-1000 | 30.2 | 0.014 | 429 | (1954) |

| Tracer | Material | Temperature range (°C) | Activation energy, Q (kcal/mole) | Frequency factor, $D_0$ (cm²/sec) | Reference No. | Year |
|---|---|---|---|---|---|---|
| **IRON OXIDE, $Fe_2O_3$** | | | | | | |
| $^{59}Fe$ | P | 760-1300 | 112 | $4 \times 10^4$ | 425 | (1952) |
| $^{59}Fe$ | P | 950-1050 | 100.2 | $1.3 \times 10^6$ | 426 | (1962) |
| **IRON OXIDE, $Fe_3O_4$** | | | | | | |
| $^{59}Fe$ | P | 750-1000 | 55.0 | 5.2 | 424 | (1953) |
| $^{59}Fe$ | P | 1000-1220 | 112 | $4 \times 10^{-5}$ | 424 | (1953) |
| $^{59}Fe$ | P | 770-1200 | 53.9 | 0.25 | 427 | (1958) |
| $^{59}Fe$ | S | 850-1075 | 84.0 | $6 \times 10^5$ | 428 | (1960) |
| $^{59}Fe$ | P | 850-1075 | 74.7 | 104 | 428 | (1960) |
| **LEAD OXIDE, PbO** | | | | | | |
| $^{212}Pb$ | P | 400-590 | 66 | $10^5$ | 429 | (1952) |
| $^{210}Pb$ | P | 600-680 | 64 | — | 430 | (1965) |
| **MAGNESIUM OXIDE, MgO** | | | | | | |
| $^{133}Ba$ | S | 1000-1725 | 77.75 | 0.07 | 441 | (1967) |
| $^{7}Be$ | S 99.99 | 1000-1700 | 36.8 | $1.41 \times 10^{-5}$ | 442 | (1966) |
| $^{45}Ca$ | S 99.99 | 900-1700 | 49.0 | $2.95 \times 10^{-5}$ | 443 | (1966) |
| Co | S | 1000-1800 | 49.4 | $5.78 \times 10^{-5}$ | 444 | (1962) |
| Fe | S | 1050-1720 | 41.6 | $8.83 \times 10^{-5}$ | 444 | (1962) |
| $^{28}Mg$ | S | 1400-1600 | 79.0 | 0.249 | 445 | (1957) |
| $^{63}Ni$ | P | 1000-1800 | 48.3 | $1.8 \times 10^{-5}$ | 446 | (1961) |
| **NICKEL OXIDE, NiO** | | | | | | |
| $^{63}Ni$ | P | 900-1000 | 55 | 0.041 | 447 | (1951) |
| $^{63}Ni$ | P | 1140-1400 | 119.5 | $2.8 \times 10^6$ | 448 | (1956) |
| $^{63}Ni$ | S | 1000-1400 | 44.2 | $3.9 \times 10^{-4}$ | 449 | (1957) |
| $^{63}Ni$ | P | 1000-1400 | 44.2 | $5.0 \times 10^{-4}$ | 449 | (1957) |
| $^{63}Ni$ | S,P | 700-1400 | 56.0 | $1.7 \times 10^{-2}$ | 450 | (1957) |
| $^{63}Ni$ | S | 1300-1700 | 53.5 | $8.01 \times 10^{-4}$ | 451 | (1960) |
| $^{63}Ni$ | S | 1000-1400 | 45.6 | $1.83 \times 10^{-3}$ | 452 | (1962) |
| $^{63}Ni$ | P 99.99 | 1200-1400 | 48.4 | $4.8 \times 10^{-4}$ | 453 | (1962) |

| Tracer | Material | Temper- ature range ($^\circ$C) | Activation energy, Q (kcal/mole) | Frequency factor, $D_0$ (cm$^2$/sec) | Reference | |
|---|---|---|---|---|---|---|
| | | | | | No. | Year |

**NIOBIUM OXIDE, $Nb_2O_5$**

| Tracer | Material | Temper- ature range | Activation energy | Frequency factor | No. | Year |
|---|---|---|---|---|---|---|
| $^{95}$Nb | P | 500-900 | 28.2 | 0.38 | 454 | (1962) |

**TIN OXIDE, $SnO_2$**

| $^{113}$Sn | P | 1000-1260 | 118.7 | $10^6$ | 455 | (1965) |
|---|---|---|---|---|---|---|

**TITANIUM OXIDE, $TiO_2$**

| $^{59}$Fe | P | 770-1000 | 55.4 | 0.192 | 416 | (1958) |
|---|---|---|---|---|---|---|
| $^{59}$Fe | P | 770-1000 | 55.0 | 0.0198 | 416 | (1958) |

**THORIUM OXIDE, $ThO_2$**

| $^{233}$Pa | | 1800-2000 | 75.4 | $2.91 \times 10^{-5}$ | 456 | (1968) |
|---|---|---|---|---|---|---|
| $^{230}$Th | | 1600-2100 | 58.8 | $1.25 \times 10^{-7}$ | 457 | (1968) |
| $^{237}$U | | 1800-2000 | 76.4 | $1.10 \times 10^{-4}$ | 458 | (1968) |

**URANIUM OXIDE, $UO_2$**

| $^{241}$Am | P | 1200-1500 | 92.0 | 0.03 | 459 | (1965) |
|---|---|---|---|---|---|---|
| $^{237}$Np | P | 1200-1500 | 109.0 | 2.9 | 459 | (1965) |
| Pa | P | 1200-1500 | 107.6 | 2.5 | 459 | (1965) |
| $^{147}$Pm | P | 1120-1410 | 56.8 | $3.5 \times 10^{-6}$ | 460 | (1961) |
| $^{239}$Pu | P | 1200-1500 | 97.3 | 0.34 | 459 | (1965) |
| Th | P | 1200-1500 | 98 | 0.16 | 459 | (1965) |
| $^{233}$U | P | 1450-1785 | 88 | $4.3 \times 10^{-4}$ | 461 | (1961) |
| $^{233}$U | P | 1300-1600 | 104.6 | 0.23 | 462 | (1961) |
| $^{233}$U | P | 1200-1500 | 104.0 | 0.9 | 459 | (1965) |
| $^{233}$U | S,P | 1450-1700 | 108.0 | 1.2 | 463 | (1965) |
| $^{233}$U | P | 1900-2150 | 72.7 | $5.82 \times 10^{-5}$ | 464 | (1966) |
| $^{91}$Y | P | 1150-1450 | 46.4 | $6.8 \times 10^{-8}$ | 461 | (1961) |
| $^{95}$Zr | P | 1120-1419 | 59.2 | $1.6 \times 10^{-6}$ | 461 | (1961) |

**YTTRIUM OXIDE, $Y_2O_3$**

| $^{91}$Y | P | 1400-1800 | 43.9 | $2.41 \times 10^{-4}$ | 465 | (1963) |
|---|---|---|---|---|---|---|

| Tracer | Material | Temperature range (°C) | Activation energy, Q (kcal/mole) | Frequency factor, $D_0$ (cm²/sec) | Reference No. | Year |
|--------|----------|------------------------|----------------------------------|-----------------------------------|---------------|------|

ZINC OXIDE, ZnO

| Tracer | Material | Temperature range (°C) | Activation energy, Q (kcal/mole) | Frequency factor, $D_0$ (cm²/sec) | Reference No. | Year |
|--------|----------|------------------------|----------------------------------|-----------------------------------|---------------|------|
| $^{65}Zn$ | P | 800-1330 | 74.0 | 1.3 | 466 | (1952) |
| $^{65}Zn$ | S | 850-940 | 20.0 | $3 \times 10^{-9}$ | 467 | (1955) |
| $^{65}Zn$ | S | 900-1025 | 73.0 | 4.8 | 468 | (1955) |
| $^{65}Zn$ | S,P | 910-1170 | 74.0 | 30 | 469 | (1956) |
| $^{65}Zn$ | P | 800-1300 | 89.0 | 0.1 | 470 | (1957) |
| $^{65}Zn$ | P | 1000-1200 | 44.0 | $1.3 \times 10^{-5}$ | 471 | (1959) |
| $^{65}Zn$ | P | 800-840 | 73.0 | 10 | 472 | (1960) |
| $^{65}Zn$ | S | 720-780 | 25.0 | $3 \times 10^{-7}$ | 473 | (1961) |

# References

1. T. S. Lundy and J. F. Murdock, J. Appl. Phys. 33(5):1671 (1962).
2. M. Beyeler and Y. Adda, J. Phys. (Paris) 29(4):345 (1968).
3. H. B. Huntington, P. B. Ghate, and J. H. Rosolowski, J. Appl. Phys. 35(10):3027 (1964).
4. A. Hassner and W. Lange, Phys. Status Solidi 11:575 (1965).
5. H. Cordes and K. Kim, Z. Naturforsch. 20a:1197 (1965); J. Appl. Phys. 37:2181 (1966).
6. J. M. Dupony, J. Mathie, and Y. Adda, Mem. Sci. Rev. Met. 63:481 (1966).
7. L. V. Pavlinov, G. V. Grigorev, and U. G. Sevastianov, Fiz. Met. i Metalloved. 25(3):565 (1968).
8. E. S. Wajda, G. A. Shirn, and H. B. Huntington, Acta Met. 3:39 (1955).
9. K. A. Mahmoud and R. Kamel, Radioisotopes in Scientific Research, Vol. 1, p. 271, Pergamon Press, New York (1958).
10. W. Hirschwald and W. Schroedter, Z. Physik. Chem. N.F. 53:392 (1967).
11. M. A. Kanter, Phys. Rev. 107:655 (1957).
12. S. Z. Bokshtein, S. T. Kishkin, and L. M. Moroz, Zavodsk. Lab. 23(3):316 (1957).
13. P. L. Gruzin, L. V. Pavlinov, and A. D. Tyutyunnik, Izv. Akad. Nauk SSSR Ser. Fiz. 5:155 (1959).
14. H. W. Paxton and E. G. Gondolf, Arch. Eisenhüttenw. 30:55 (1959).
15. S. Z. Bokshtein, S. T. Kishkin, and L. M. Moroz, Investigation of the Structure of Metals by Radioactive Isotope Methods, State Publishing House of the Ministry of Defense Industry, Moscow (1959); AEC-tr-4505 (1961).
16. N. A. Bogdanov, Russ. Met. Fuels (English transl.) 3:95 (1960).
17. L. I. Ivanov, M. P. Matveeva, V. A. Morozov, and D. A. Prokoshkin, Russ. Met. Fuels (English transl.) 2:63 (1962).
18. W. C. Hagel, Trans. Met. Soc. AIME 224(3):430 (1962).
19. J. Askill and D. H. Tomlin, Phil. Mag. 11(111):467 (1965).
20. F. C. Nix and F. E. Jaumot, Phys. Rev. 82:72 (1951).
21. R. C. Ruder and C. E. Birchenall, Trans. Met. Soc. AIME 191:142 (1951).
22. P. L. Gruzin, Dokl. Akad. Nauk SSSR 86:289 (1952).
23. H. W. Mead and C. E. Birchenall, Trans. Met. Soc. AIME 203(9):994 (1955).
24. K. Hirano, R. P. Agarwala, B. L. Averbach, and M. Cohen, J. Appl. Phys. 33(10):3049 (1962).

25.     A. Hassner and W. Lange, Phys. Status Solidi 8:77 (1965).

26.     B. V. Rollin, Phys. Rev. 55:231 (1939).

27.     J. Steigman, W. Shockley, and F. C. Nix, Phys. Rev. 56:13 (1939).

28.     M. S. Maier and H. R. Nelson, Trans. Met. Soc. AIME 147:39 (1942).

29.     C. L. Raynor, L. Thomassen, and J. Rousse, Trans. ASM 30:313 (1942).

30.     A. Kuper, H. Letaw, L. Slifkin, E. Sonder, and C. T. Tomizuka, Phys. Rev.
        96:1224 (1954); errata, ibid. 98:1870 (1956).

31.     W. L. Mercer, Ph.D. Thesis, Leeds University, England (1955).

32.     K. Monma, H. Suto, and H. Oikawa, J. Japan Inst. Metals 28:188 (1964).

33.     E. A. Smirnov, L. I. Ivanov, and E. A. Abranyan, Izv. Akad. Nauk SSSR
        Metal., 168 (1967).

34.     H. Letaw, L. M. Slifkin, and W. M. Portnoy, Phys. Rev. 93(4):892 (1954).

35.     H. Letaw, W. M. Portnoy, and L. Slifkin, Phys. Rev. 102:636 (1956).

36.     H. Widner and G. R. Gunther-Mohr, Helv. Phys. Acta 34(6):635 (1961).

37.     H. A. C. McKay, Trans. Faraday Soc. 34:845 (1938).

38.     H. C. Gatos and A. Azzam, Trans. Met. Soc. AIME 194:407 (1952).

39.     H. C. Gatos and A. D. Kurtz, Trans. Met. Soc. AIME 200:616 (1954).

40.     A. D. Kurtz, B. L. Averbach, and M. Cohen, Acta Met. 3:442 (1955).

41.     B. Okkerse, Phys. Rev. 103(5):1246 (1956).

42.     H. W. Mead and C. E. Birchenall, Trans. Met. Soc. AIME 209(7):874 (1957).

43.     S. M. Makin, A. H. Rowe, and A. D. LeClaire, Proc. Phys. Soc. B70(6):545
        (1957).

44.     D. Duhl, K. Hirano, and M. Cohen, Acta Met. 11(1):1 (1963).

45.     H. M. Gilder and D. Lazarus, J. Phys. Chem. Solids 26:2081 (1965).

46.     T. F. Archbold and W. H. King, Trans. Met. Soc. AIME 233(4):839 (1965).

47.     F. R. Winslow and T. S. Lundy, Trans. Met. Soc. AIME 233:1790 (1965).

48.     R. E. Eckert and H. G. Drickamer, J. Chem. Phys. 20:13 (1952).

49.     J. E. Dickey, Acta Met. 7:350 (1959).

50.     J. Jagielak, G. Pawlicki, and A. Sobaszek, Nukleonika 10(a):541 (1965).

51.     C. E. Birchenall and R. F. Mehl, J. Appl. Phys. 19:217 (1948).

52.     C. E. Birchenall and R. F. Mehl, Trans. Met. Soc. AIME 188:144 (1950).

53.     F. S. Buffington, I. D. Bakalar, and M. Cohen, Physics of Powder Metallurgy,
        p. 92, McGraw-Hill, New York (1951).

54.     P. L. Gruzin, Probl. Metalloved. i Fiz. Met. 3:201 (1952).

55.     A. A. Zhukhovitskii and V. A. Geodakyan, Primenenie Radioaktivn. Izotopov
        v Metallurg. Sb. 34:267 (1955); AEC-tr-3100, Uses of Radioactive Isotopes in
        Metallurgy Symposium XXXIV, Pt. 2, p. 52.

56.     V. M. Golikov and V. T. Borisov, Problems of Metals and Physics of Metals,
        4th Symposium (1955), Consultants Bureau, New York (1957); AEC-tr-2924
        (1958).

57.     C. Leymonie and P. Lacombe, Rev. Met. (Paris) 55:524 (1958); Compt. Rend.
        245:1922 (1957); Metaux (Corrosion-Ind.) 34:457 (1959); Metaux (Corrosion-
        Ind.) 35:45 (1960).

58.     R. J. Borg and C. E. Birchenall, Trans. Met. Soc. AIME 218:980 (1960).

59.     F. S. Buffington, K. Hirano, and M. Cohen, Acta Met. 9(5):434 (1961).

60.     D. Graham and D. H. Tomlin, Phil. Mag. 8:1581 (1963).

61.    V. M. Amonenko, A. M. Blinkin, and I. G. Ivantsov, Fiz. Metal. i Met-
       taloved. 17(1):56 (1964); Phys. Metals Metallog. (USSR) (English transl.)
       17(1):54 (1964).
62.    D. W. James and G. M. Leak, Phil. Mag. 14:701 (1966).
63.    P. L. Gruzin, Izv. Akad. Nauk SSSR Otd. Tekhn. Nauk 3:383 (1953).
64.    H. W. Mead and C. E. Birchenall, Trans. Met. Soc. AIME 206:1336 (1956).
65.    S. Z. Bokshtein, S. T. Kishkin, and L. M. Moroz, Metalloved. i Term.
       Obrabotka Metal. 2:2 (1957).
66.    S. D. Gertsriken and M. P. Pryanishnikov, Vopr. Fiz. Met. i Metalloved.,
       Sb. Nauchn. Rabot. Inst. Metallofiz. Akad. Nauk Ukr. SSR 9:147 (1958).
67.    N. A. Bogdanov, Russ. Met. Fuels (English transl.) 2:61 (1962).
68.    B. Sparke, D. W. James, and G. M. Leak, J. Iron Steel Inst. (London)
       203(2):152 (1965).
69.    L. I. Staffansson and C. E. Birchenall, AFOSR-733 (1961).
70.    R. J. Borg, D. Y. F. Lai, and O. Krikorian, Acta Met. 11(8):867 (1963).
71.    G. von Hevesy, W. Seith, and A. Keil, Z. Physik 79:197 (1932); Z. Metallk.
       25:104 (1935).
72.    B. Okkerse, Acta Met. 2:551 (1954).
73.    N. H. Nachtrieb and G. S. Handler, J. Chem. Phys. 23(9):1569 (1955).
74.    H. A. Resing and N. H. Nachtrieb, J. Phys. Chem. Solids 21:40 (1961).
75.    P. G. Shewmon and F. N. Rhines, Trans. Met. Soc. AIME 200:1021 (1954).
76.    P. G. Shewmon, Trans. Met. Soc. AIME 206:918 (1956).
77.    Y. V. Borisov, P. L. Gruzin, and L. V. Pavlinov, Met. i Metalloved.
       Chistykh Metal. 1:213 (1959); translated in JPRS-5195.
78.    M. B. Bronfin, S. Z. Bokshtein, and A. A. Zhukhovitskii, Zavodsk. Lab.
       26(7):828 (1960).
79.    W. Danneberg and E. Krautz, Z. Naturforsch. 16a(a):854 (1961).
80.    J. Askill and D. H. Tomlin, Phil. Mag. 8(90):997 (1963).
81.    L. V. Pavlinov and V. N. Bikov, Fiz. Met. i Metalloved. 18:459 (1964).
82.    R. E. Hoffman, F. W. Pikus, and R. A. Ward, Trans. Met. Soc. AIME
       206:483 (1956).
83.    J. E. Reynolds, B. L. Averbach, and M. Cohen, Acta Met. 5:29 (1957).
84.    W. R. Upthegrove and M. J. Sinnott, Trans. ASM 50:1031 (1958).
85.    J. R. MacEwan, J. U. MacEwan, and L. Yaffe, Can. J. Chem. 37(10):1623
       (1959).
86.    J. R. MacEwan, J. U. MacEwan, and L. Yaffe, Can. J. Chem. 37(10):1629
       (1959).
87.    A. Messner, R. Benson, and J. E. Dorn, Trans. ASM 53:227 (1961).
88.    A. R. Wazzam, J. Mote, and J. E. Dorn, U.S. Report AD-257188 (1961).
89.    A. Y. Shinyaev, Phys. Metals Metallog. (USSR) (English transl.) 15(1):100
       (1963).
90.    G. B. Federov, E. A. Smirnov, and F. I. Zhomov, Met. i Metalloved.
       Chistykh Metal., Sb. Nauchn. Rabot 4:110 (1963).
91.    A. R. Wazzam, J. Appl. Phys. 36(11):3596 (1965).
92.    A. R. Wazzam and J. E. Dorn, J. Appl. Phys. 36(1):222 (1965).
93.    I. G. Ivanov, Fiz. Metal. i Metalloved. 22(5):725 (1966).

94.    H. Baker, Phys. Stat. Solidi 28(2):596 (1968).

95.    D. F. Kalinovich, I. I. Kovenskii, and M. D. Smolin, Fiz. Tverd. Tela
       10(2):569 (1968).

96.    R. Resnick and L. S. Castleman, Trans. Met. Soc. AIME 218:307 (1960).

97.    R. F. Peart, D. Graham, and D. H. Tomlin, Acta Met. 10:519 (1962).   '

98.    T. S. Lundy, F. R. Winslow, R. E. Pawel, and C. J. McHargue, Trans. Met.
       Soc. AIME 233:1533 (1965).

99.    V.D.Lyubimov,P. V. Geld, and G. P. Shveykin, Izv. Akad. Nauk SSSR Met.
       i Gorn. Delo 5:137 (1964); Russ. Met. Mining 5:100 (1964).

100.   N. L. Peterson, Phys. Rev. 136(2A):A568 (1964).

101.   N. H. Nachtrieb and G. S. Handler, J. Chem. Phys. 23:1187 (1955).

102.   G. V. Kidson and R. Ross, Proc. 1st UNESCO Int. Conf. Radioisotopes in
       Scientific Res., p. 185 (1958).

103.   F. Cattaneo, E. Germagnoli, and F. Grasso, Phil. Mag. 7:1373 (1962).

104.   J. N. Mundy, L. W. Barr, and F. A. Smith, Phil. Mag. 15(134):411 (1967).

105.   R. E. Tate and G. R. Edwards, Symposium on Thermodynamics with Emphasis
       on Nuclear Materials and Atomic Solids, pp. 105-113 in IAEA Symposium
       Vol. II, International Atomic Energy Agency, Vienna (1966).

106.   R. E. Tate and E. M. Cramer, Trans. Met. Soc. AIME 230:639 (1964).

107.   M. Dupuy and D. Calais, AIME 242:1679 (1968).

108.   B. I. Boltaks and B. T. Plachenov, Soviet Phys. — Tech. Phys. (English transl.)
       27(10):2071 (1957).

109.   R. F. Peart, Phys. Status Solidi 15:K119 (1966).

110.   B. J. Masters and J. M. Fairfield, Appl. Phys. Letters 8:280 (1966).

111.   W. A. Johnson, Trans. Met. Soc. AIME 143:107 (1941).

112.   D. Turnbull, Phys. Rev. 76:471 (1949).

113.   R. E. Hoffman and D. Turnbull, J. Appl. Phys. 22:634 (1951); errata p. 984.

114.   L. Slifkin, D. Lazarus, and C. T. Tomizuka, J. Appl. Phys. 23:1032 (1952).

115.   R. D. Johnson and A. B. Martin, Phys. Rev. 86:642 (1952); NAA-SR-170
       (1952).

116.   S. N. Kriukov and A. A. Zhukhovitskii, Dokl. Akad. Nauk SSSR 90:379
       (1953).

117.   B. N. Finkelshtein and A. I. Yamashchikova, Dokl. Akad. Nauk SSSR 98:781
       (1954).

118.   A. A. Zhukhovitskii and V. A. Geodakyan, Dokl. Akad. Nauk SSSR 102:301
       (1955); AEC-tr-2265 (1955).

119.   S. D. Gertsriken and M. P. Pryanishnikov, Vopr. Fiz. Met. i Metalloved.,
       Sb. Nauchn. Rabot. Inst. Metallofiz. Akad. Nauk Ukr. SSR 4:528 (1955).

120.   E. Sonder, Phys. Rev. 100:1662 (1955).

121.   H. Kruegar and H. N. Hirsch, Trans. Met. Soc. AIME 203:125 (1955).

122.   C. T. Tomizuka and E. Sonder, Phys. Rev. 103(5):1182 (1956).

123.   V. P. Vasilev, Dokl. Akad. Nauk SSSR 110:61 (1956); AEC-tr-2972.

124.   N. H. Nachtrieb, J. Petit, and J. Wehrenberg, J. Chem. Phys. 26(1):106
       (1957).

125.   S. D. Gertsriken and T. K. Yatsenko, Vopr. Fiz. Met. i Metalloved., Sb.
       Nauchn. Rabot. Inst. Metallofiz. Akad. Nauk Ukr. SSR. 8:101 (1957).

126. M. E. Yanitskaya, A. A. Zhukhvitskii, and S. Z. Bokshtein, Dokl. Akad. Nauk SSSR 112:720 (1957).

127. G. Airoldi and E. Germagnoli, Energia Nucl. (Milan) 5:445 (1958).

128. S. D. Gertsriken and D. D. Tsitsiliano, Phys. Metals Metallog. (USSR) (English transl.) 6:80 (1958).

129. A. V. Savitskii, Fiz. Metal. i Metalloved. 10:564 (1960); A. V. Savitskii, Phys. Metals Metallog. (USSR) (English transl.) 10:71 (1960).

130. A. V. Savitskii, Fiz. Metal. i Metalloved. 16:886 (1963).

131. J. Kucera, Czech.J. Phys. B14:915 (1964).

132. Y. Imai and T. Miyazaki, Sci. Rep. Res. Inst. Thoku Univ. Ser A 2:59 (1966).

133. V. N. Kaigorodov, S. M. Klotsman, A. N. Timofeev, and I. Sh. Trakhtenberg, Fiz. Met. i Metalloved. 25(5):910 (1968).

134. P. Reimer, Metall. 22(6):577 (1968).

135. N. H. Nachtrieb, E. Catalano, and J. A. Weil, J. Chem. Phys. 20(8):1185 (1952).

136. J. N. Mundy, L. W. Barr, and F. A. Smith, Phil. Mag. 14:785 (1966).

137. R. B. Cuddeback and H. G. Drickamer, J. Chem. Phys. 19:790 (1951).

138. R. N. Ghoshtagore, Phys. Rev. 155(3):698 (1967).

139. R. L. Eager and D. B. Langmuir, Phys. Rev. 89:911 (1953); errata p. 890; D. B. Langmuir, Phys. Rev. 86:642 (1952).

140. P. L. Gruzin and V. I. Meshkov, Vopr. Fiz. Met. i Metalloved., Sb. Nauchn. Rabot. Inst. Metallofiz. Akad. Nauk Ukr. SSR 570 (1955); AEC-tr-2926.

141. R. E. Pawel and T. S. Lundy, J. Phys. Chem. Solids 26:937 (1965).

142. G. A. Shirn, Acta Met. 3:87 (1955).

143. F. Schmitz and M. Fock, J. Nucl. Materials 21:317 (1967).

144. W. Boas and P. J. Fensham, Nature 164:1127 (1949).

145. P. J. Fensham, Australian J. Sci. Research, Ser. A, A3:91 (1950); errata A4:229 (1951).

146. J. D. Meakin and E. Klokholm, Trans. Met. Soc. AIME 218(3):463 (1960).

147. W. Chomba and J. Andruszkiewicz, Nukleonika 5(10):611 (1960).

148. W. Lange and A. Hassner, Phys. Status Solidi 1:50 (1961).

149. C. Coston and N. H. Nachtrieb, J. Phys. Chem. 68:2219 (1964).

150. G. Pawlicki, Nukleonika 12(12):1123 (1967).

151. C. M. Libanati and S. F. Dyment, Acta Met. 11:1263 (1963).

152. J. F. Murdock, T. S. Lundy, and E. E. Stansbury, Acta Met. 12(9):1033 (1964).

153. N. E. W. DeReca and C. M. Libanati, Acta Met. 16(10):1297 (1968).

154. V. P. Vasilev and S. G. Chernomorchenko, Zavodsk. Lab. 22:688 (1956); AEC-tr-4276.

155. W. Danneberg, Metall. 15:977 (1961).

156. R. L. Andelin, J. D. Knight, and M. Kahn, Trans. Met. Soc. AIME 233:19 (1965).

157. G. M. Neumann and W. Hirschwald, Z. Naturforsch. 21a:812 (1966).

158. Y. Adda, A. Kirianenko, and C. Mairy, Compt. Rend. 253:445 (1961); J. Nucl. Mater. 6(1):130 (1962).

159. Y. Adda, A. Kirianenko, and C. Mairy, J. Nucl. Mater. 1(3):300 (1959).

160.   A. A. Bochvar, V. G. Kuznetsova, and V. S. Sergeev, Proc. 2nd U.N. Intern.
       Conf. Peaceful Uses At. Energy, Geneva, 1958, 6:8 (1958).
161.   Y. Adda and A. Kirianenko, Compt. Rend. 247(9):744 (1958).
162.   S. J. Rothman, L. T. Lloyd, and A. L. Harkness, Trans. Met. Soc. AIME
       218(4):605 (1960).
163.   T. S. Lundy and C. J. McHargue, Trans. Met. Soc. AIME 233:243 (1965).
164.   R. F. Peart, J. Phys. Chem. Solids 26:1853 (1965).
165.   R. P. Agarwala, S. P. Murarka, and M. S. Anand, Acta Met. 16(1):61 (1968).
166.   F. R. Banks, Phys. Rev. 59:376 (1941).
167.   P. H. Miller and F. R. Banks, Phys. Rev. 61:648 (1942).
168.   H. B. Huntington, G. A. Shirn, and E. S. Wajda, Phys. Rev. 87:211 (1952).
169.   G. A. Shirn, E. S. Wajda, and H. B. Huntington, Acta Met. 1:513 (1953).
170.   T. Liu and H. G. Drickamer, J. Chem. Phys. 22:314 (1954).
171.   F. E. Jaumot and R. L. Smith, Trans. Met. Soc. AIME 206:137 (1956).
172.   I. A. Naskidashvili and V. M. Dolidze, Soobshch. Akad. Nauk Gruz. 18:671
       (1957).
173.   J. E. Hilliard, B. L. Averbach, and M. Cohen, Acta Met. 7:86 (1959).
174.   N. L. Peterson and S. J. Rothman, Phys. Rev. 163(3):645 (1967).
175.   P. L. Gruzin, V. S. Emelyanov, G. G. Ryabova, and G. B. Federov, Proc.
       2nd U.N. Intern. Conf. Peaceful Uses At. Energy, 1958, 19:187 (1958).
176.   V. S. Lyashenko, B. N. Bikov, and L. V. Pavlinov, Phys. Metals Metallog.
       (USSR) (English transl.) 8(3):40 (1959).
177.   P. Flubacher, E.I.R. Bericht No. 49, May (1963).
178.   G. B. Federov and V. D. Gulyakin, Met. i Metalloved. Chistykh Metal.
       1:170 (1959).
179.   Y. V. Borisov, Yu. G. Godin, and P. L. Gruzin, Metallurgy and Metallography,
       Part I, NP-tr-448, p. 196 (1960).
180.   D. Yolokoff, S. May, and Y. Adda, Compt. Rend. 251(3):2341 (1960).
181.   G. V. Kidson and J. McGurn, Can. J. Phys. 39(8):1146 (1961).
182.   J. I. Federer and T. S. Lundy, Trans. Met. Soc. AIME 227:592 (1963)
183.   M. S. Anand and R. P. Agarwala, Trans. AIME 239(11):1848 (1967).
184.   S. Badrinarayanan, and H. B. Mathers, Int. J. Appl. Radiat. Isotopes 19(4):353
       (1968).
185.   M. S. Anand, S. P. Murarka, and R. P. Agarwala, J. Appl. Phys. 36(12):3860
       (1965).
186.   K. Hirano, R. P. Agarwala, and M. Cohen, Acta Met. 10(9):857 (1962).
187.   R. P. Agarwala, S. P. Murarka, and M. S. Anand, Acta Met. 12(8):871 (1964).
188.   A. R. Paul and R. P. Agarwala, J. Appl. Phys. 38(9):3790 (1967).
189.   S. P. Murarka, M. S. Anand, and R. P. Agarwala, Acta Met. 16(1):69 (1968).
190.   D. E. Ovsienko and I. K. Zasimchuh, Fiz. Metal. i Metalloved. 10:743
       (1960).
191.   M. C. Naik, J. M. Dupony, and Y. Adda, Soc. France de Met. Autumn Conf.
       (1964).
192.   M. C. Naik, J. M. Dupony, and Y. Adda, Mem. Sci. Rev. Met. 63:488 (1966).
193.   J. R. Wolfe, D. R. McKenzie, and R. J. Borg, J. Appl. Phys. 36(6):1906 (1965).
194.   H. K. Lonsdale and J. N. Graves, J. Appl. Phys. 38(9):3620 (1967).

195.   H. W. Paxton and R. A. Wolfe, Trans. AIME 230:1426 (1964).

196.   I. I. Kovenski, Fiz. Metal. i Metalloved. 16:613 (1963).

197.   M. Aucouturier, M. O. Pinheiro, Rib De Castro, and P. Lacombe, Acta Met. 13:125 (1965).

198.   M. M. Pavlyuchenko and I. F. Kononyuk, Dokl. Akad. Nauk Belorusskoi SSR 8:157 (1964).

199.   C. T. Tomizuka, cited by D. Lazarus, Solid State Phys. Vol. 10 (1960).

200.   A. B. Martin and F. Asaro, Phys. Rev. 80:123 (1950); A. B. Martin, R. D. Johnson, and F. Asaro, J. Appl. Phys. 25:364 (1954).

201.   T. Hirone, N. Kunitomi, M. Sakamoto, and H. Yamaki, J. Phys. Soc. Japan 13(8):838 (1958).

202.   M. Sakamoto, J. Phys. Soc. Japan 13(8):845 (1958).

203.   C. A. Machliet, Phys. Rev. 109(6):1964 (1958).

204.   Y. Tomono and A. Ikushima, J. Phys. Soc. Japan 13(8):762 (1958).

205.   J. G. Mullen, Phys. Rev. 121:1649 (1961).

206.   A. Ikushima, J. Phys. Soc. Japan 14:111 (1959).

207.   S. Yukawa and M. J. Sinott, Trans. AIME 203:996 (1955).

208.   A. Ikushima, J. Phys. Soc. Japan 14(11):1636 (1959).

209.   H. P. Bonzel, Z. Elektrochem. 70:73 (1966).

210.   K. Monma, H. Suto, and H. Oikawa, Nippon Kinsoku Gakkaishi 28:192 (1964).

211.   H. P. Bonzel, Acta Met. 13:1084 (1965).

212.   K. J. Anusavice, J. J. Pinajian, H. Oikawa, and R. T. DeHoff, Trans. AIME 242(9):2027 (1968).

213.   N. L. Peterson, Phys. Rev. 132(6):2471 (1963).

214.   R. D. Johnson and B. H. Faulkenberry, ASD-TDR-63-625 (July 1963).

215.   M. C. Inman and L. W. Barr, Acta Met. 8(2):112 (1960).

216.   P. P. Kuzmenko, L. F. Ostrovskii, and V. S. Kovalchuk, Fiz. Tverd. Tela 4:490 (1962).

217.   S. Komura and N. Kunitomi, J. Phys. Soc. Japan 18 (Suppl. 2):208 (1963).

218.   J. Hino, C. T. Tomizuka, and C. Wert, Acta Met. 5(1):41 (1957).

219.   R. T. DeHoff, A. G. Guy, K. J. Anusavice, and T. B. Lindemer, Trans AIME 236:881 (1966).

220.   V. E. Kosenko, Fiz. Tverd. Tela 1:1622 (1959); Soviet Phys. — Solid State (English transl.) 1:1481 (1959).

221.   A. A. Bugai, V. E. Kosenko, and E. G. Miselynuk, Zh. Tekhn. Fiz. 27(1):207 (1957); NP-tr-448, p. 219 (1960).

222.   A. V. Sandulova, M. I. Droniuk, and V. M. P'dak, Fiz. Tverd. Tela 3:2913 (1961).

223.   R. C. Miller and F. M. Smith, Phys. Rev. 107(1):65 (1957).

224.   B. I. Boltaks, V. P. Grabtchak, and T. D. Dzafarov, Fiz. Tverd. Tela 6:3181 (1964).

225.   V. D. Ignatkov and V. E. Kosenko, Fiz. Tverd. Tela 4(6):1627 (1962); Soviet Phys. — Solid State (English transl.) 4(6):1193 (1962).

226.   V. I. Tagirov and A. A. Kuliev, Fiz. Tverd. Tela 4(1):272 (1962); Soviet Phys. — Solid State (English transl.) 4(1):196 (1962).

227.    W. C. Mallard, A. B. Gardner, R. F. Bass, and L. M. Slifkin, Phys. Rev.
        129(2):617 (1963).

228.    S. M. Klotsman, N. K. Arkhipova, A. N. Timofeyev, and L. S. Trakhtenberg,
        Fiz. Metal. i Metalloved. 20(3):390 (1965).

230.    A. J. Mortlock, A. H. Rowe, and A. D. LeClaire, Phil. Mag. 5:803 (1960).

231.    T. R. Anthony and D. Turnbull, Phys. Rev. 151:495 (1966).

232.    R. J. Borg and D. Y. F. Lai, Acta Met. 11(8):861 (1963).

233.    P. L. Gruzin, V. G. Kostogonov, and P. A. Platonov, Probl. Metalloved. i
        Fiz. Metal. 4:517 (1955).

234.    P. L. Gruzin, Yu. A. Polikarpov, G. B. Federov, and M. A. Shumilov, Metal-
        lurgy and Metallography, NP-Tr-448, p. 119 (1960).

235.    C. G. Homan, Acta Met. 12:1071 (1964).

236.    P. L. Gruzin and D. F. Litvin, Dokl. Akad. Nauk SSSR 94(1):41 (1954).

237.    P. L. Gruzin, Dokl. Akad. Nauk SSSR 94:681 (1954).

238.    V. W. Lange, A. Hassner, and E. Dahn, Neue Hütte 1:33 (1961).

239.    K. Sato, Trans. Japan Inst. Metals 5:91 (1964).

240.    A. M. Huntz, M. Aucouturier, and P. Lacombe, C. R. Acad. Sci. Paris Ser. C
        265(10):554 (1967).

241.    M. S. Anand and R. P. Agarwala, J. Appl. Phys. 37:4248 (1966).

242.    A. V. Tomilov and G. V. Shcherbedinskii, Fiz. Khim. Met. Mater. 3(3):261
        (1967).

243.    V. T. Borisov, V. M. Golikov, and G. V. Shcherbedinskii, Fiz. Metal. i
        Metalloved. 22(1):159 (1966).

244.    K. Hirano, M. Cohen, and B. L. Averbach, Acta Met. 9(5):440 (1961).

245.    P. L. Gruzin and V. V. Mural, Fiz. Metal. i Metalloved. 16(4):551 (1963);
        Phys. Metals Metallog. (USSR) (English transl.) 16(4):50 (1963).

246.    G. Seibel, Compt. Rend. 256(22):4661 (1963).

247.    G. Bruggeman and J. Roberts, J. Met. 20(8):54 (1968).

248.    J. Kieszniewski, Pr. Inst. Hutn. 19(4):253 (1967).

249.    G. V. Grigorev and L. V. Pavlinov, Fiz. Metal. i Metalloved. 25(5):836
        (1968).

250.    P. L. Gruzin and D. F. Litvin, Probl. Metalloved. i Fiz. Met. 4:486 (1955).

251.    T. Suzuoka, Japan. Inst. Metals 2:176 (1961).

252.    R. Lindner and F. Karnik, Acta Met. 3:297 (1955).

253.    G. Seibel, Compt. Rend. 255(23):3182 (1962).

254.    B. F. Dyson, T. Anthony, and D. Turnbull, J. Appl. Phys. 37:2370 (1966).

255.    A. Ascoli, J. Inst. Metals 89(6):218 (1960).

256.    G. V. Kidson, Phil. Mag. 13:247 (1966).

257.    J. W. Miller, Am. Phys. Soc. Bull. 12(7):1072 (1967).

258.    H. A. Resing and N. H. Nachtrieb, Phys. Chem. Solids 21(1/2):40 (1961).

259.    J. N. Mundy, A. Ott, and L. Lowenberg, Z. Naturforsch. 22a(12):2113 (1967).

260.    A. Ott and A. Norden-Ott, Z. Naturforsch. 23(3):473 (1968).

261.    A. Y. Nakonechnikov, L. V. Pavlinov, and V. N. Bikov, Fiz. Metal. i
        Metalloved. 22:234 (1966).

262.    R. F. Peart, D. Graham, and D. H. Tomlin, Acta Met. 10:519 (1962).

263.    J. Askill, Phys. Status Solidi 9(2):K113 (1965).

264. G. P. Benediktova, G. N. Dubinin, M. G. Kapman, and G. V. Shcherbedinskii, Metalloved. i Term. Obrabotka Metal. 5:55 (1966).

265. S. Z. Bokshtein, M. B. Bronfin, and S. T. Kishkin, Diffusion Processes, Structure, and Properties of Metals, pp. 16, 24, Consultants Bureau, New York (1965).

266. B. A. Vandyshev and A. S. Panov, Fiz. Metal. i Metalloved. 25(2):321 (1968).

267. L. V. Pavlinov, A. Y. Nakonechnikov, and V. N. Bikov, At. Energ. (USSR) 19:521 (1965).

268. J. Askill, Phys. Status Solidi 23:K21 (1967).

269. L. M. Larikov, Svoistva i Primenenie Zharoprochnykh Splavov, Akad. Nauk SSSR Inst. Metallurgii 28 (1966).

270. A. Chatterjee and D. J. Fabian, J. Inst. Metals 96(6):186 (1968).

271. P. L. Gruzin, Yu. A. Polikarpov, and G. B. Federov, Fiz. Metal. i Metalloved. 4(1):94 (1957); Phys. Metals Metallog. USSR (English transl.) 4(1):74 (1957).

272. R. C. Ruder and C. E. Birchenall, Trans. AIME 191:142 (1951).

273. S. P. Murarka, M. S. Anand, and R. P. Agarwala, J. Appl. Phys. 35(4):1339 (1964).

274. M. B. Neiman, A. I. Chinaev, and B. G. Dzantiev, Dokl. Akad. Nauk SSSR 91:265 (1953).

275. V. P. Vasilev, I. F. Kamardin, V. I. Skatskii, S. G. Chernomorchenko, and G. N. Shuppe, Trudy Sred. Gos. Uni. im V. I. Lenina 65:47 (1955); Translation AEC-tr-4272.

276. P. Lacombe, Colloque sur les Joints de Grains, Saclay, 1960, Presses Universitaires de France, Paris, 1961.

277. I. G. Ivanov, Fiz. Metal. i Metalloved. 22(5):725 (1966).

278. P. P. Kuzmenko and G. P. Grinevich, Fiz. Tverd. Tela 4(11):3266 (1962); Soviet Phys. — Solid State (English transl.) 4(11):2390 (1962).

279. S. T. Kishkin and S. Z. Bokshtein, Proc. Intern. Conf. Peaceful Uses At. Energy, Geneva, 1955, 15:81 (1956).

280. P. P. Kuzmenko and G. Grinevich, Fiz. Metal. i Metalloved. 24(3):424 (1967).

281. H. W. Allison and G. E. Moore, J. Appl. Phys. 29(5):842 (1958).

282. P. V. Geld and V. D. Lyubimov, Izv. Akad. Nauk SSSR OTN Met. i Toplivo 6:119 (1961).

283. P. Son, S. Ihara, M. Miyake, and T. Sano, J. Japan. Inst. Met. 31:998 (1967).

284. J. Pelleg, J. Metals 20(8):54 (1968).

285. V. D. Lyubimov, P. V. Geld, G. P. Shveikin, and Yu. A. Sutina, Izv. Akad. Nauk SSSR, Metal. 2:84 (1967).

286. B. A. Vandyshev and A. S. Panov, Izv. Akad. Nauk SSSR 1:206 (1968).

287. J. Askill, Phys. Status Solidi 9(3):K167 (1965).

288. L. W. Barr, J. N. Mundy, and F. A. Smith, Phil. Mag. 16:1139 (1967).

289. A. A. Kuliev and D. N. Nasledov, Zh. Tekhn. Fiz. 28(2):259 (1958); Soviet Phys. — Tech. Phys. (English transl.) 3(2):235 (1958).

290. J. D. Struthers, J. Appl. Phys. 27(12):1560 (1956); errata 28(4):516 (1957).

291. W. R. Wilcox and T. J. LaChapelle, J. Appl. Phys. 35(1):240 (1964).

292.    R. C. Newman and J. Wakefield, Phys. Chem. Solids 19(3):230 (1961).

293.    B. I. Boltaks and I. I. Sozinov, Zh. Tekhn. Fiz. 28(3):679 (1958); Soviet
        Phys. — Tech. Phys. (English transl.) 3(3):636 (1958).

294.    H. P. Bonzel, Phys. Status Solidi 20:493 (1967).

295.    S. Maekawa, J. Phys. Soc. Japan 17(10):1592 (1962).

296.    J. J. Rahan, M. E. Pickering, and J. Kennedy, J. Electrochem. Soc. 106(8):705
        (1959).

297.    F. E. Jaumot and A. Sawatskii, J. Appl. Phys. 27:1186 (1956).

298.    C. T. Tomizuka and L. M. Slifkin, Phys. Rev. 96:610 (1954).

299.    T. Hirone and H. Yamamoto, J. Phys. Soc. Japan 16(3):455 (1961).

300.    K. Hirano, M. Cohen, and B. L. Averbach, Acta Met. 11:463 (1963).

301.    A. Sawatskii and F. E. Jaumot, Trans. AIME 209:1207 (1957).

302.    R. E. Hoffman, Acta Met. 6:95 (1958).

303.    V. N. Kaigorodov, Ya. A. Rabovskii, and V. K. Talinskii, Fiz. Metal. i
        Metalloved. 24(1):117 (1967).

304.    T. Hirone, S. Miura, and T. Suzuoka, J. Phys. Soc. Japan 16(12):2456 (1961).

305.    J. H. Holloman and D. Turnbull, SO-2030 (1953).

306.    R. E. Hoffman, D. Turnbull, and E. W. Hart, Acta Met. 3:417 (1955).

307.    C. B. Pierce and D. Lazarus, Phys. Rev. 114:686 (1959).

308.    N. Barbouth, J. Ouder, and J. Cabane, C. R. Acad. Sci. Paris Ser. C
        264(12):1029 (1967).

309.    L. M. Slifkin, D. Lazarus, and C. T. Tomizuka, J. Appl. Phys. 23:1405
        (1952).

310.    E. Sonder, L. M. Slifkin, and C. T. Tomizuka, Phys. Rev. 93(5):970 (1954).

311.    V. N. Kaigorodov, Ya. A. Rabovskii, and V. K. Talinskii, Fiz. Metal. i
        Metalloved. 24(4):661 (1967).

312.    A. Sawatskii and F. E. Jaumot, Phys. Rev. 100:1627 (1955).

313.    N. L. Peterson and S. J. Rothman, Phys. Rev. 154(3):558 (1967).

314.    D. F. Kalinovich, I. I. Kovenskii, and M. D. Smolin, Fiz. Metal. i Metalloved.
        18:314 (1964).

315.    P. Son, S. Ihara, M. Miyake, and T. Sano, J. Japan. Inst. Met. 30(12):1137
        (1966).

316.    Sh. Movalanov and A. A. Kuliev, Fiz. Tverd. Tela 4(2):542 (1962).

317.    N. I. Ibraginov, M. G. Shachtachtinskii, and A. A. Kuliev, Fiz. Tverd.
        Tela 4:3321 (1962).

318.    T. R. Anthony, B. F. Dyson, and D. Turnbull, J. Appl. Phys. 39(3):1391
        (1968).

319.    B. F. Dyson, J. Appl. Phys. 37:2375 (1966).

320.    Z. Rozycki and A. Sobaszek, Nukleonika 12(9):677 (1967).

321.    W. Chomka and J. Andruszkiewicz, Nukleonika 5(10):611 (1960).

322.    A. Sawatskii, J. Appl. Phys. 29(9):1303 (1958).

323.    I. I. Kovenskii, Ukr. Fiz. Zh. 8:797 (1963).

324.    A. J. Mortlock and D. H. Tomlin, Phil. Mag. 4(4):628 (1959).

325.    G. B. Gibbs, D. Graham, and D. H. Tomlin, Phil. Mag. 8(92):1269 (1963).

326.    R. F. Peart and D. H. Tomlin, Acta Met. 10:123 (1962).

327.    L. V. Pavlinov, Fiz. Metal. i Metalloved. 24(2):272 (1967).

328.  J. Askill and G. B. Gibbs, Phys. Status Solidi 11:557 (1965).

329.  S. Z. Bokshtein, S. T. Kishkin, and V. B. Osvenskii, Metalloved. i Term. Obrabotka Metal. 6:21 (1960); Metal Sci. Heat Treat. Metals (USSR) (English transl.) 4/6:329 (1960).

330.  J. F. Murdock, T. S. Lundy, and E. E. Stansbury, Acta Met. 12(9):1033 (1964).

331.  J. F. Murdock and C. J. McHargue, Acta Met. 16(4):493 (1968).

332.  L. N. Larikov, V. M. Tyshkevich, and L. F. Chorna, Ukr. Fiz. Zh. 12(6):983 (1967).

333.  S. J. Rothman, J. Nucl. Mater. 3(1):77 (1961).

334.  N. L. Peterson and S. J. Rothman, Phys. Rev. 136(3A):A842 (1964).

335.  S. J. Rothman, ANL-6320 (1961).

336.  J. H. Rosolowski, Phys. Rev. 124(6):1828 (1961).

337.  P. B. Ghate, Phys. Rev. 130(1):174 (1963).

338.  A. P. Batra and H. B. Huntington, Phys. Rev. 145:542 (1966).

339.  N. A. Pleteneva and N. P. Fedoseeva, Dokl. Akad. Nauk SSSR 151(2):384 (1963).

340.  A. P. Batra and H. B. Huntington, Phys. Rev. 154(3):569 (1967).

341.  R. P. Agarwala, S. P. Murarka, and M. S. Anand, Trans. Met. Soc. AIME 233:986 (1965).

342.  A. M. Blinkin and V. V. Vorobiov, Ukr. Fiz. Zh. 9(1):91 (1964).

343.  R. P. Agarwala and A. R. Paul, Proc. Nucl. Rad. Chem. Symp., Poona, India (March, 1967), p. 542.

344.  Y. V. Borisov, Y. G. Godin, P. L. Gruzin, A. I. Evstyukhin, and V. S. Yemelyanov, Met. i Met., Izdatel. Akad. Nauk SSSR, Moscow, 1958; Translation NP-TR-448, p. 196 (1960).

345.  L. V. Pavlinov and V. N. Bikov, Fiz. Metal. i Metalloved. 19(3):397 (1965).

346.  R. A. Andriewski, V. N. Zagryazkin, and G. Ya. Meshcheryakov, Fiz. Metal. i Metalloved. 21:140 (1966).

347.  L. V. Pavlinov, Fiz. Metal. i Metalloved. 24(2):272 (1967).

348.  T. S. Lundy and J. I. Federer, Trans. AIME 227:592 (1963).

349.  T. Heumann and H. Bohmer, J. Phys. Chem. Solids 29(2):237 (1968).

350.  A. E. Berkowitz, F. E. Jaumot, Jr., and F. C. Nix, Phys. Rev. 95:1185 (1954).

351.  L. I. Ivanov and N. P. Ivanchev, Izv. Akad. Nauk SSSR, Otd. Tekhn. Nauk, No. 8, p. 15 (1958).

352.  H. W. Paxton and T. Kunitake, Trans. Met. Soc. AIME 218:1003 (1960).

353.  A. Ya. Shinyayev, Fiz. Metal. i Metalloved. 20:875 (1965).

354.  A. Ya. Shinyayev, Metallurgy and Metallography (1958); NP-Tr-448, p. 210 (1960).

355.  S. D. Gertsricken and I. Y. Dekhtyar, Proc. U.N. Intern. Conf. Peaceful Uses At. Energy Conf., Geneva, 15:99-102 (1955), United Nations, New York (1956).

356.  S. D. Gertsricken and I. Y. Dekhtyar, Fiz. Metal. i Metalloved. 3:242 (1956); Phys. Metals Metallog. USSR (English transl.) 3:242 (1956).

357.  P. L. Gruzin and B. M. Noskov, Problems of Metallography and the Physics of Metals, Fourth Symposium, ed. B. Ya. Lyubov, AEC-Tr-2924, pp. 355-360 (1955).

358.   T. Hirone, N. Kunitomi, and M. Sakamoto, J. Phys. Soc. Japan 13:840 (1958).
359.   I. I. Kovenskii, Fiz. Tverd. Tela 3(2):350 (1961); Soviet Phys. – Solid State
       (English transl.) 3(2):252 (1961).
360.   R. Ebeling and H. Wever, Z. Metallk. 53(3):222 (1968).
361.   M. C. Inman, D. Johnston, W. L. Mercer, and R. Shuttleworth, p. 85 in
       Radioisotope Conference, 1954, Vol. II, Physical Sciences and Industrial
       Applications, ed. J. E. Johnston, Butterworths, London (1954).
362.   A. B. Kuper, D. Lazarus, J. R. Manning, and C. T. Tomizuka, Phys. Rev.
       104:1536 (1956).
363.   W. C. Mallard, A. B. Gardner, R. F. Bass, and L. M. Slifkin, Phys. Rev.
       129:617 (1963).
364.   H. B. Huntington, N. C. Miller, and V. Nerses, Acta Met. 9:749 (1961).
365.   D. Gupta, D. Lazarus, and D. S. Lieberman, Phys. Rev. 153(3):558 (1967).
366.   G. V. Meshcheryakov, R. A. Andrievskii, and V. N. Zagryazkin, Fiz. Metal.
       i Metalloved. 25(1):189 (1968).
367.   S. D. Gertsricken, I. Y. Dekhtyar, L. M. Kumok, and E. G. Madatova,
       Metallurgy and Metallography, NP-Tr-448, p. 130 (1960).
368.   L. V. Pavlinov, E. A. Isadzanov, and V. P. Smirnov, Fiz. Metal. i Metalloved.
       25(5):959 (1968).
369.   R. A. Wolfe and H. W. Paxton, Trans. Met. Soc. AIME 230:1426 (1964).
370.   R. Lindner and F. Karnik, Acta Met. 3:297 (1955).
371.   S. D. Gertsricken and M. P. Pryanishnikov, Ukr. Fiz. Zh. 3:255 (1958).
372.   I. G. Ivanov and A. M. Blinkin, Fiz. Metal. i Metalloved. 19:274 (1965).
373.   P. L. Gruzin, Yu. V. Kornev, and G. V. Kurdyumov, Dokl. Akad. Nauk SSSR
       80:49 (1951).
374.   H. W. Mead and C. E. Birchenall, Trans. AIME 206:1336 (1956).
375.   V. Linnenbom, M. Tetenbaum, and C. Cheek, J. Appl. Phys. 26:932 (1955).
376.   A. F. Smith and G. B. Gibbs, Metal Sci. J. 2(2):47 (1968).
377.   P. L. Gruzin, B. M. Noskov, and V. I. Shirokov, Problems of Metallography
       and the Physics of Metals, Fourth Symposium, ed. B. Ya. Lyubov, AEC-Tr-
       2924, pp. 350-354 (1955).
378.   M. B. Neiman and A. Shinyayev, Dokl. Akad. Nauk SSSR 102:969 (1955).
379.   S. Z. Bokshtein, V. A. Kazakova, S. T. Kishkin, and L. M. Mirskii, Izv.
       Akad. Nauk SSSR, Otd. Tekhn. Nauk, Energ. i Avtomat., No. 12, p. 18
       (1955).
380.   P. L. Gruzin and E. V. Kuznetsov, Problems of Metallography and the Physics
       of Metals, Fourth Symposium, ed. B. Ya. Lyubov, AEC-Tr-2924, pp. 346-349
       (1955).
381.   B. Mills, G. K. Walker, and G. M. Leak, Phil. Mag. 12:939 (1965).
382.   H. Schenck and K. W. Lange, Archiv. Eisenhüttenw. 37(10):809 (1966).
383.   D. Y. F. Lai and R. J. Borg, Trans. Met. Soc. AIME 233:1973 (1965).
384.   I. K. Kupalova and S. V. Zemskii, Metalloved. Term. Obrabotka Metal.
       2:10 (1968).
385.   H. A. Domain and H. I. Aaronson, Trans. Met. Soc. AIME 230:44 (1964).
386.   D. F. Kalinovich, I. I. Kovenskii, and M. D. Smolin, Fiz. Metal. i Metalloved.
       11(2):307 (1961); Phys. Metals Metallog. USSR (English transl.) 11(2):148
       (1961).

387.    I. N. Frantsevich, D. F. Kalinovich, I. I. Kovenskii, and M. D. Smolin, Ukr.
        Fiz. Zh. 8(9):1020 (1963).
388.    M. D. Smolin, Fiz. Metal. i Metalloved. 15(3):472 (1963); Phys. Metals
        Metallog. USSR (English transl.) 15(3):131 (1963).
389.    D. F. Kalinovich, I. I. Kovenskii, and M. D. Smolin, Ukr. Fiz. Zh. 10:1365
        (1965).
390.    S. D. Gertsricken, I. Ya. Dekhtyar, and V. S. Mikhalenkov, Metallurgy and
        Metallography, NP-Tr-448, p. 176 (1960).
391.    P. L. Gruzin and G. B. Federov, Dokl. Akad. Nauk SSSR 105:264 (1955).
392.    A. Ya. Shinyayev, Metallurgy and Metallography, NP-Tr-448, p. 154.
393.    A. D. Tyutyunnik and G. V. Estulin, Fiz. Metal. i Metalloved. 4:558 (1957);
        Phys. Metals Metallog. USSR (English transl.) 4(3):146 (1957).
394.    S. D. Gertsricken, I. Ya. Dekhtyar, and L. M. Kumok, Vopr. Fiz. Metal. i
        Metalloved., Akad. Nauk Ukr. SSR, Sb. Nauchn. Rabot, No. 5, p. 71 (1954).
395.    E. W. DeReca and C. Pampillo, Acta Met. 15(8):1263 (1967).
396.    A. Ya. Shinyaev, Izv. Akad. Nauk SSSR, Metal.1:203 (1968).
397.    P. V. Geld, Z. Ya. Velmozhnyi, V.D.Lyubimov,and G. P. Shveykin, Izv.
        Vysshikh Uchebn. Zavedenii Tsvetn. Met. 2:135 (1966).
398.    G. B. Gibbs, D. Graham, and D. H. Tomlin, Phil. Mag. 8:1269 (1962).
399.    S. D. Gertsricken and T. K. Yatsenko, Vopr. Fiz. Metal. i Metalloved.,
        Akad. Nauk Ukr. SSR, Sb. Nauchn. Rabot, No. 8, p. 101 (1957).
400.    Alan Schoen, Self-Diffusion in Alpha Solid Solutions of Silver–Cadmium
        and Silver–Indium, Convair, General Dynamics Corp. Div., Scientific Re-
        search Lab. Research Report No. 2 (March 1958).
401.    N. H. Nachtrieb, J. Petit, and J. Wehrenberg, J. Chem. Phys. 26(1):106
        (1957).
402.    S. D. Gertsricken and D. D. Tsitsiliano, Nauk. Pov. K.D.U. Fiz. Vol. 1
        (1956).
403.    E. Sonder, Phys. Rev. 100:1662 (1955).
404.    M. Butsyk and S. Gertsricken, Zh. Tekhn. Fiz. 20:428 (1950).
405.    D. Lazarus and C. T. Tomizuka, Phys. Rev. 103:1155 (1956).
406.    R. E. Hoffman and D. Turnbull, J. Appl. Phys. 23:1409 (1952).
407.    N. V. Grum-Grzhimailo, Izv. Akad. Nauk SSSR, Otd. Tekhn. Nauk, No. 7,
        p. 24 (1957).
408.    P. M. Arzhany, R. M. Volkova, and D. A. Prokoshkin, Izv. Akad. Nauk SSSR,
        Otd. Tekhn. Nauk, Met. i Toplivo, No. 6, p. 162 (1962).
409.    A. Adda, A. Kirianenko, and M. Bendazzoli, Compt. Rend. 253:653 (1961).
410.    C. J. Santoro, Am. Phys. Soc. Bull. 12(7):1041 (1967).
411.    I. A. Naskidashvili, Avtoreferat kandidatskoi dissertatsii, Tbilisi, Akad. Nauk
        Gruz. SSSR (1957).
412.    V. S. Lyashenko, V. N. Bykov, and L. V. Pavlinov, Fiz. Metal. i Metalloved.
        8(3):362 (1959); Phys. Metals Metallog. USSR (English transl.) 8(3):40 (1959).
413.    A. E. Paladino and W. D. Kingery, J. Chem. Phys. 37:957 (1962).
414.    W. Fiedler and D. Bobleter, Atomkernenergie 13(1):57 (1968).
415.    D. V. Ignatov, I. N. Belokurova, and I. N. Belyanin, Metallurgy and Metal
        Research (All-Union Scientific and Technical Conference on the Use of

Radioactive and Stable Isotopes and Radiation in the National Economy and in Science), Izd. Akad. Nauk SSSR, Moscow (1958), p. 326; NP-tr-448, p. 256 (1958).

416.   V. I. Izvekov and K. M. Gorbunova, Fiz. Metal. i Metalloved. 7:713 (1959).
417.   V. O. Bobleter, W. Fiedler, and F. Grasso, Atomkemenergie 10:261 (1965).
418.   R. W. Redington, Phys. Rev. 87:1066 (1952).
419.   S. B. Austerman and D. G. Swarthout, NAA-SR-5893 (May 1961).
420.   S. B. Austerman, J. Nucl. Mater. 14:248 (1964).
421.   H. J. de Bruin and G. M. Watson, J. Nucl. Mater. 14:239 (1964); ORNL-3526 (1964).
422.   S. B. Austerman and J. W. Wagner, J. Am. Ceram. Soc. 49:94 (1966).
423.   H. J. de Bruin, G. M. Watson, and C. M. Blood, J. Appl. Phys. 37:4573 (1966).
424.   G. D. Palkar, D. N. Sitharamarao, and A. K. Dasgupta, Trans. Faraday Soc. 59:2634 (1963).
425.   R. Lindner, S. Austrumdal, and A. Akerstrom, Acta Chem. Scand. 6:468 (1952).
426.   Y. P. Gupta and L. J. Wevrick, J. Phys. Chem. Solids 28:811 (1967).
427.   R. Lindner and A. Akerstrom, Z. Physik. Chem. 6:162 (1956).
428.   W. C. Hagel and A. U. Seybolt, J. Electrochem. Soc. 108:1146 (1961).
429.   R. E. Carter and F. D. Richardson, J. Metals 6:1244 (1954).
430.   A. I. Andrievskii, A. V. Sandulova, and M. I. Yurkevich, Fiz. Tverd. Tela 2:624 (1960); Soviet Phys. — Solid State (English transl.) 2:581 (1960).
431.   W. J. Moore and B. Selikson, J. Chem. Phys. 19:1539 (1951); 20:927 (1952).
432.   A. V. Sandulova, M. I. Dronyuk, V. M. Rybak, and K. S. Shcherbai, Ukr. Fiz. Zh. 7:289 (1962).
433.   A. V. Sandulova and D. Ee-Tsin, Fiz. Tverd. Tela, 2(5):874 (1960).
434.   L. Himmel, N. F. Mehl, and C. E. Birchenall, J. Metals 5:827 (1953).
435.   R. Lindner, Arkiv. Kemi 4:381 (1952).
436.   V. I. Izvekov, N. S. Gorbukov, and A. A. Babad-Zachryapin, Fiz. Metal. i Metalloved. 14:195 (1962).
437.   V. I. Izvekov, Inzh. Fiz. Zh. Akad. Nauk Belorussk. SSR 1:64 (1958).
438.   S. M. Klotsmann, A. N. Timofeyev, and I. Sh. Trakhtenberg, Fiz. Metal. i Metalloved. 10:93 (1960).
439.   R. Lindner, Arkiv Kemi 4:385 (1952); 7:273 (1954).
440.   A. K. Dasgupta, D. N. Sitharamarao, and G. D. Palkar, Nature 207(4997):628 (1965).
441.   B. C. Harding, Phil. Mag. 16:1039 (1967).
442.   B. C. Harding and A. J. Mortlock, J. Chem. Phys. 45(7):2699 (1966).
443.   J. Rungis and A. J. Mortlock, Phil. Mag. 14(130):821 (1966).
444.   B. J. Wuensch and T. Vasilos, J. Chem. Phys. 36(11):2917 (1962).
445.   R. Lindner and G. D. Parfitt, J. Chem. Phys. 26:182 (1957).
446.   B. J. Wuensch and T. Vasilos, Reactivity of Solids, p. 57, Elsevier, New York (1961).
447.   W. J. Moore, Complete Scientific Report on Contract Termination [on] Rate Processes in Inorganic Solids at High Temperatures [for] October 1, 1950, to August 1, 1951, ORO-78 (1951).

448.    R. Lindner and A. Akerstrom, Z. Physik. Chem. 6:162 (1956).

449.    M. T. Shim and W. J. Moore, J. Chem. Phys. 26:802 (1957).

450.    R. Lindner and A. Akerstrom, Disc. Faraday Soc. 23:133 (1957).

451.    W. J. Moore, J. J. Landers, S. R. Logan, M. O'Keeffe, J. S. Choi, J. Ebisuzaki,
        S. Brown, and D. Mitchell, Physical Chemistry of the Solid State, TID-11020
        (1960).

452.    J. S. Choi and W. J. Moore, J. Phys. Chem. 66:1308 (1962).

453.    S. M. Klotsmann, A. N. Timofeyev, and I. Sh. Traкhtenberg, Fiz. Metal. i
        Metalloved. 14:428 (1962).

454.    D. L. Douglass, Proc. Conf. Corrosion Reactor Materials, Salzburg, 1962, 2:223
        (1962), IAEA.

455.    R. Lindner and O. Enqvist, Arkiv Kemi 9:471 (1956).

456.    H. Furuya and S. Yajima, J. Nucl. Mater. 25(1):38 (1968).

457.    R. J. Hawkins and C. B. Alcock, J. Nucl. Mater. 26(1):112 (1968).

458.    H. Furuya, J. Nucl. Mater. 26(1):123 (1968).

459.    F. Schmitz and R. Lindner, J. Nucl. Mater. 17:259 (1965).

460.    F. Schmitz and R. Lindner, Z. Naturforsch. 16a:1096 (1961).

461.    A. B. Austern and J. Belle, J. Nucl. Mater. 3:267 (1961).

462.    R. Lindner and F. Schmitz, Z. Naturforsch. 16a:1373 (1961).

463.    C. B. Alcock, R. J. Hawkins, A. W. D. Hills, and P. McNamara, Symposium
        on Thermodynamics with Emphasis on Nuclear Materials and Atomic Trans-
        port in Solids, Vienna, 1965.

464.    S. Yajima, H. Furuya, and T. Hirai, Sci. Rep. Res. Inst. Tohoku Univ.
        18(Suppl.):238 (1966); S. Yajima, H. Furuya, and T. Hirai, J. Nucl. Mater.
        20(2):162 (1966).

465.    M. F. Berand and D. R. Wilder, J. Appl. Phys. 34:2318 (1963).

466.    R. Lindner, Acta Chem. Scand. 6:457 (1952).

467.    F. Munnich, Naturwissenschaften 42:340 (1955).

468.    E. A. Secco and W. J. Moore, J. Chem. Phys. 23:1170 (1955).

469.    E. Spicar, "Untersuchung über Einstellung des Fehlstellen-Gleichgewichtes
        in Verbindungskristallen," Thesis, University of Stuttgart (1956).

470.    J. P. Roberts and C. Wheeler, Phil. Mag. 2:708 (1957).

471.    W. J. Moore and E. L. Williams, Disc. Faraday Soc. 28:86 (1959).

472.    E. A. Secco, Disc. Faraday Soc. 28:94 (1959).

473.    E. A. Secco, Can. J. Chem. 39:1544 (1961).

# Author Index

# General Index